Autodesk Fusion 360 Exercises
Learn by Practicing (2023-24)

Design 100 Real-World 3D Models by Practicing

CADArtifex

The premium provider of learning products and solutions
www.cadartifex.com

Autodesk Fusion 360 Exercises - Learn by Practicing (2023-24)
Author: Sandeep Dogra
Email: info@cadartifex.com

Published by
CADArtifex
www.cadartifex.com

Copyright © 2022 CADArtifex

NOTICE TO THE READER

Examination Copies

Electronic Files

Disclaimer

www.cadartifex.com

Dedication

First and foremost, I would like to thank my parents for being a great support throughout my career and while writing this book.

Heartfelt gratitude goes to my wife and my sisters for their patience and endurance in supporting me to take up and successfully accomplish this challenge.

I would also like to acknowledge the efforts of the employees at CADArtifex for their dedication in editing the contents of this book.

Table of Contents

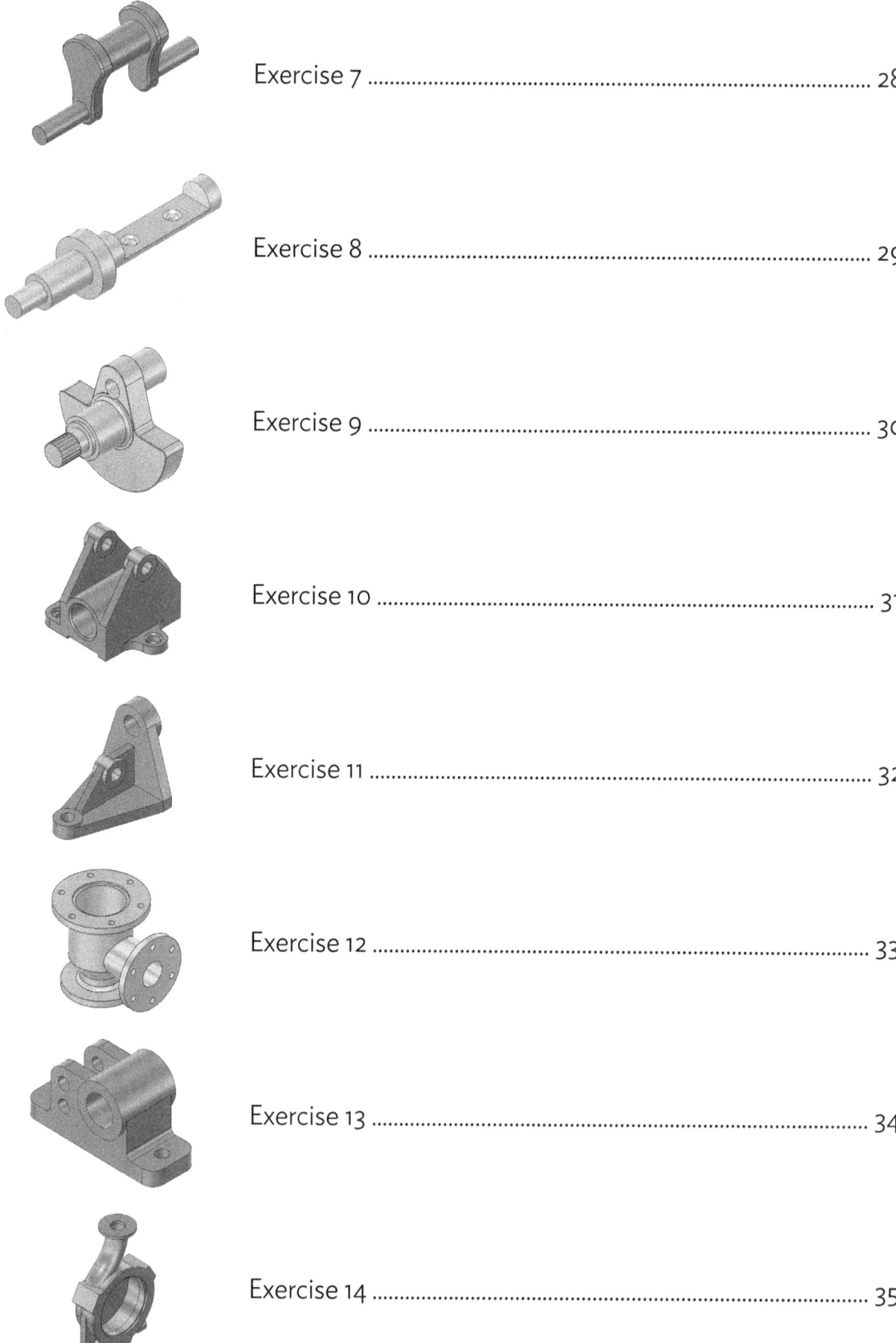

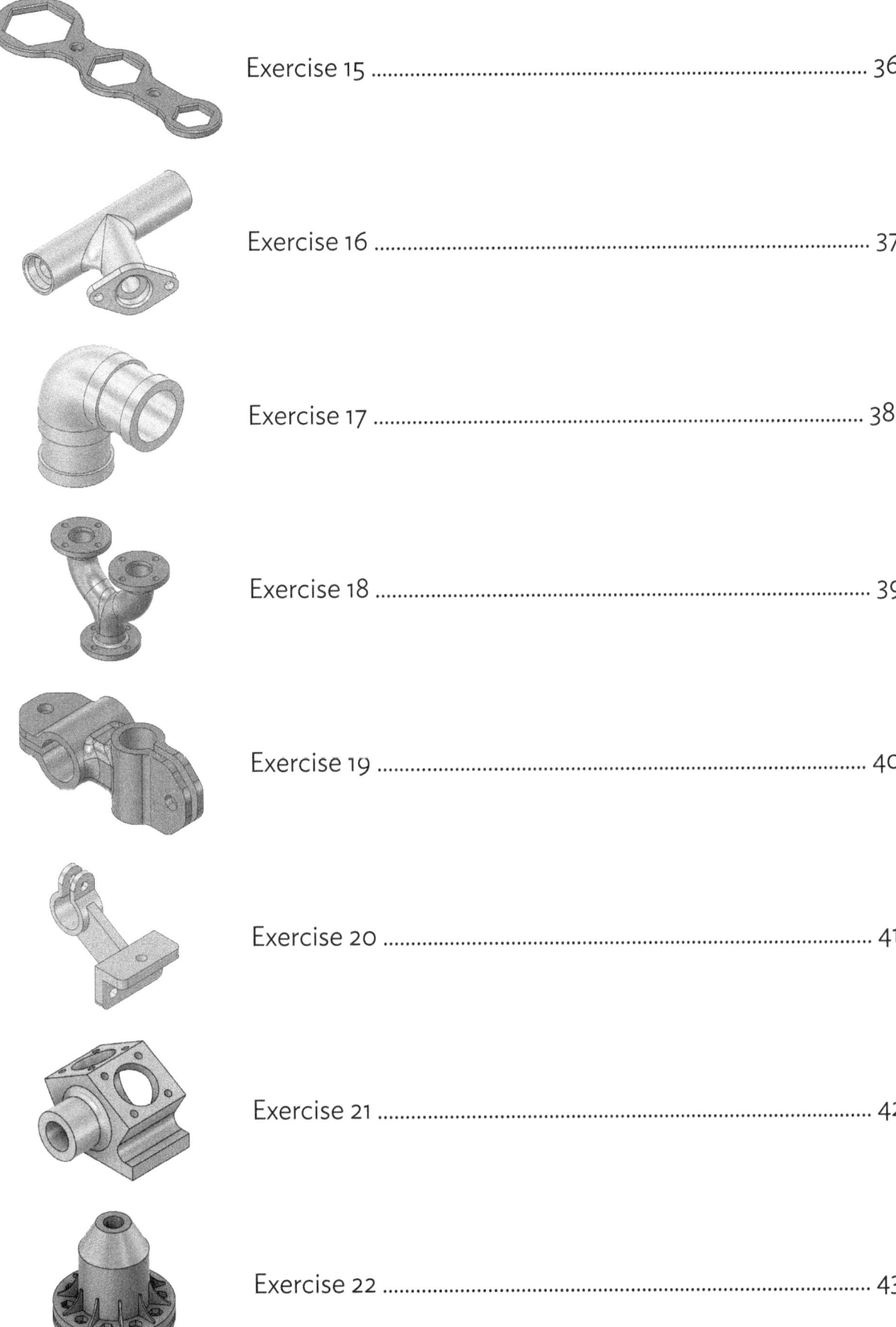

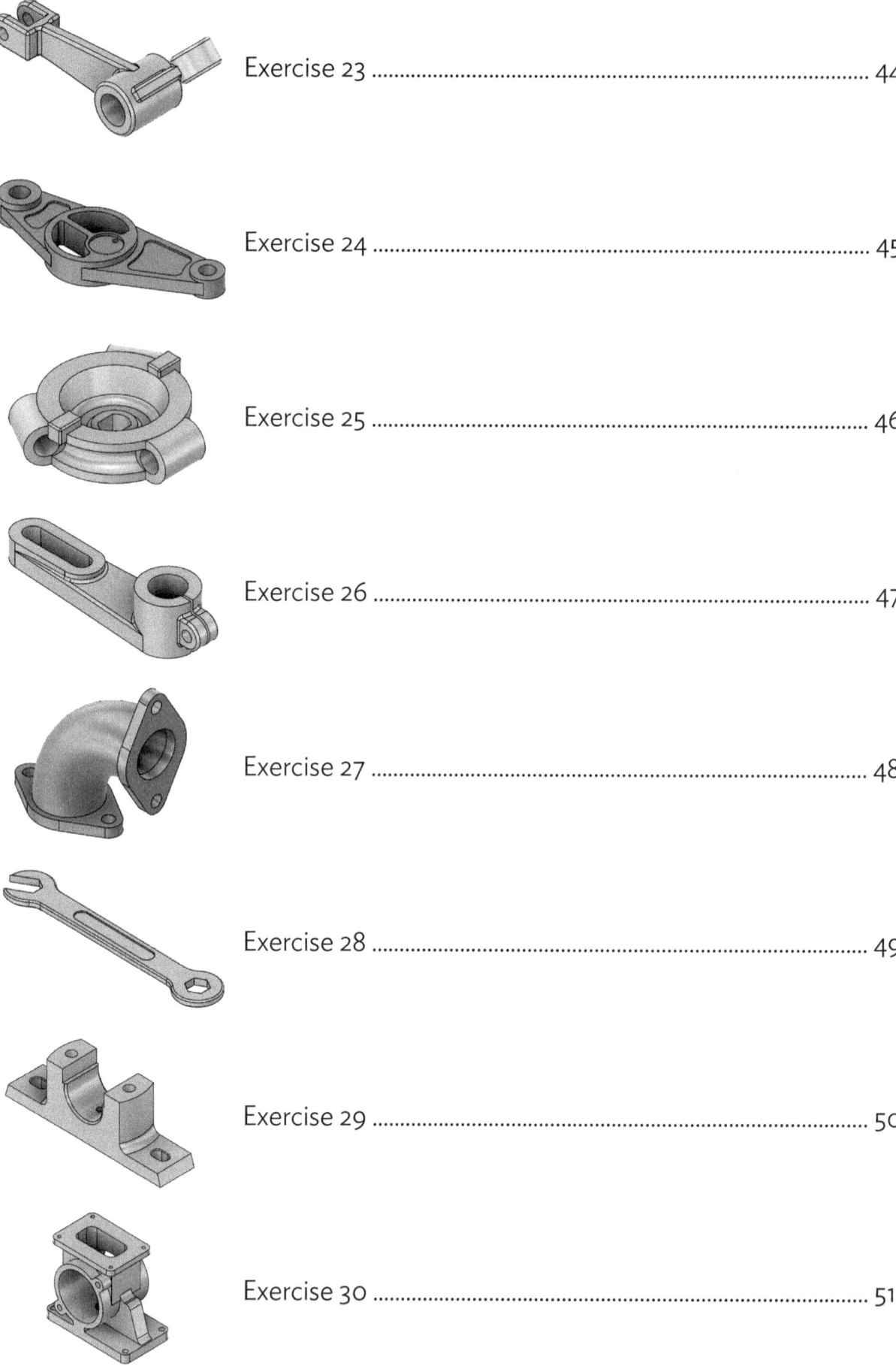

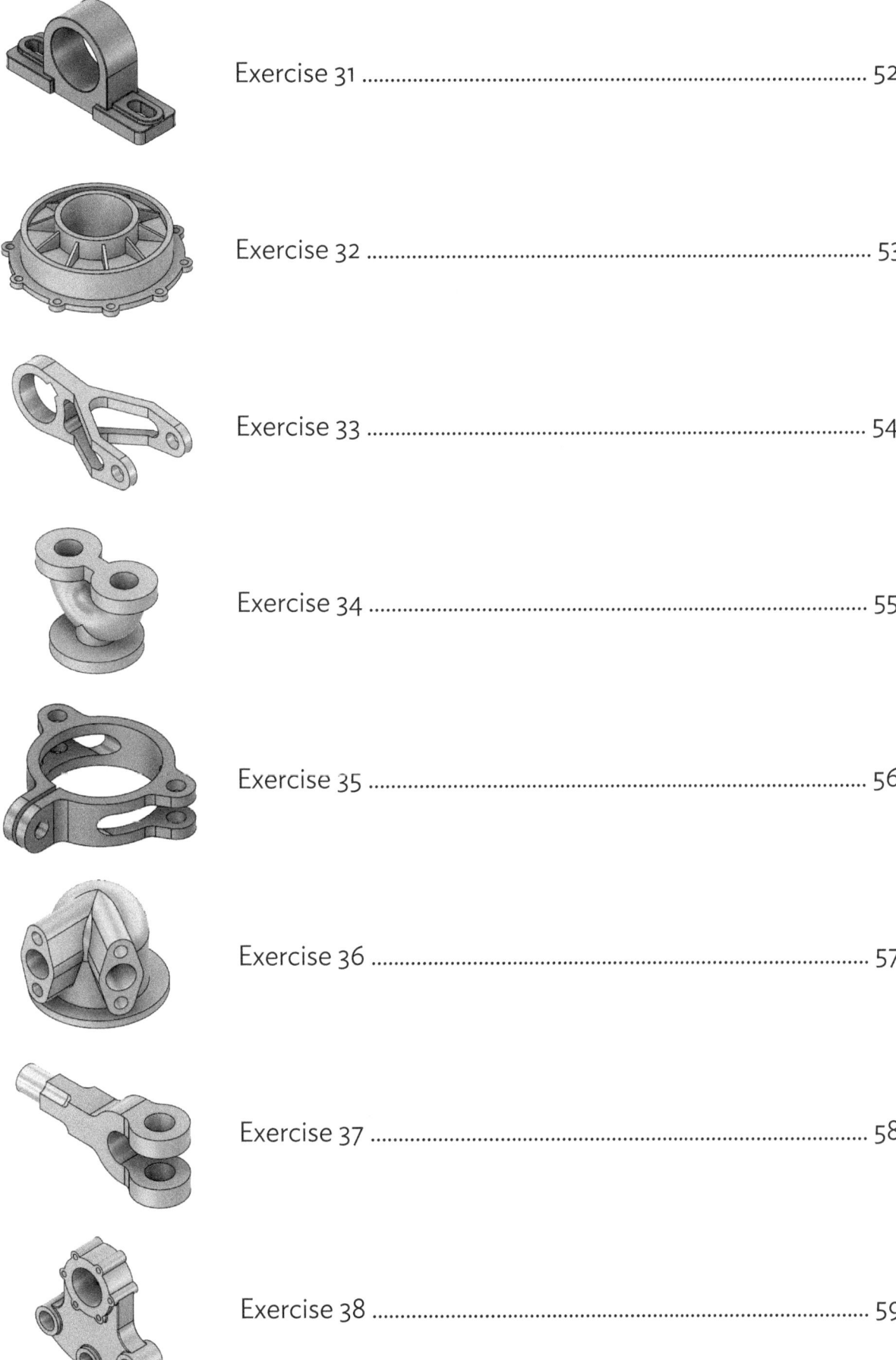

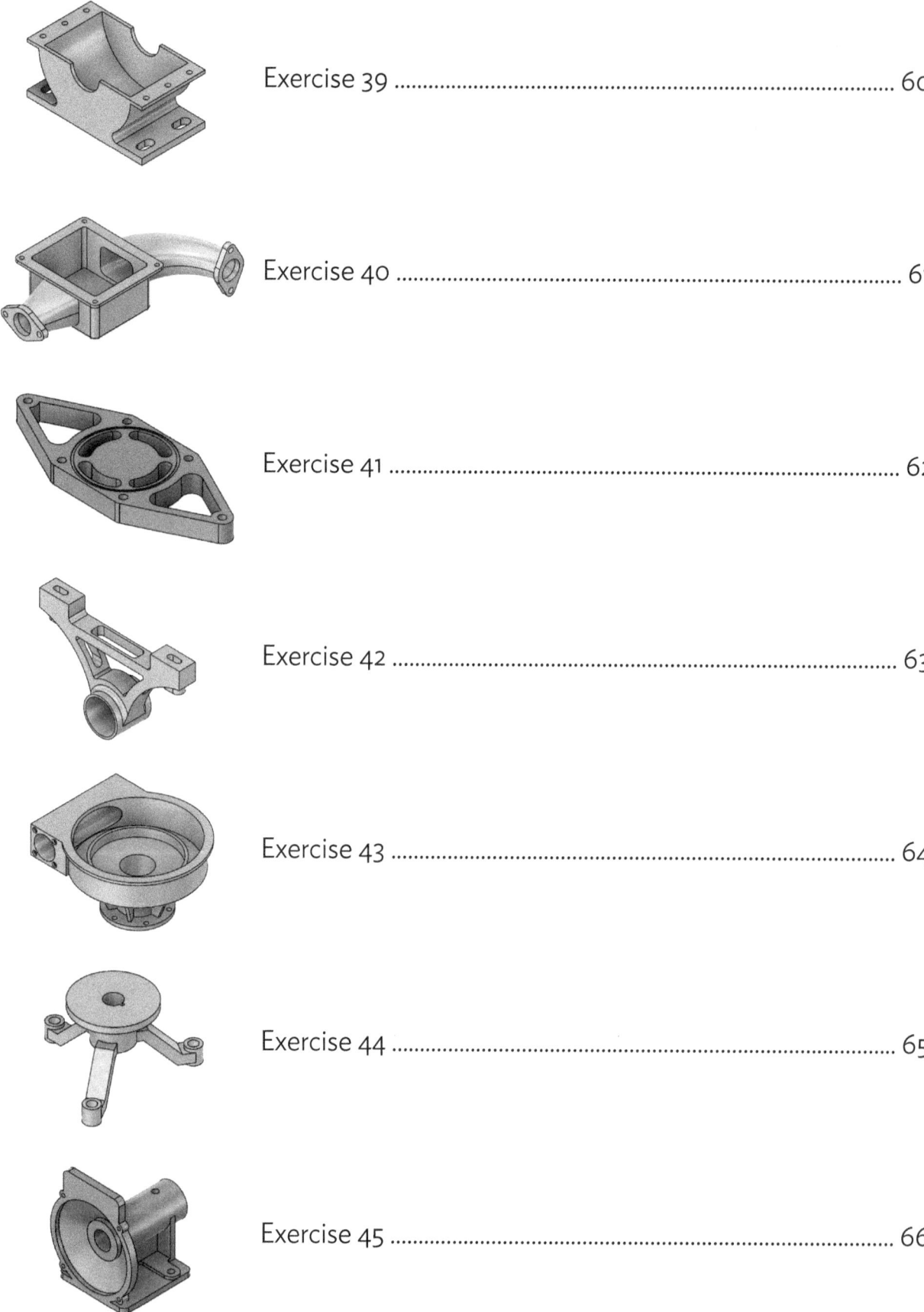

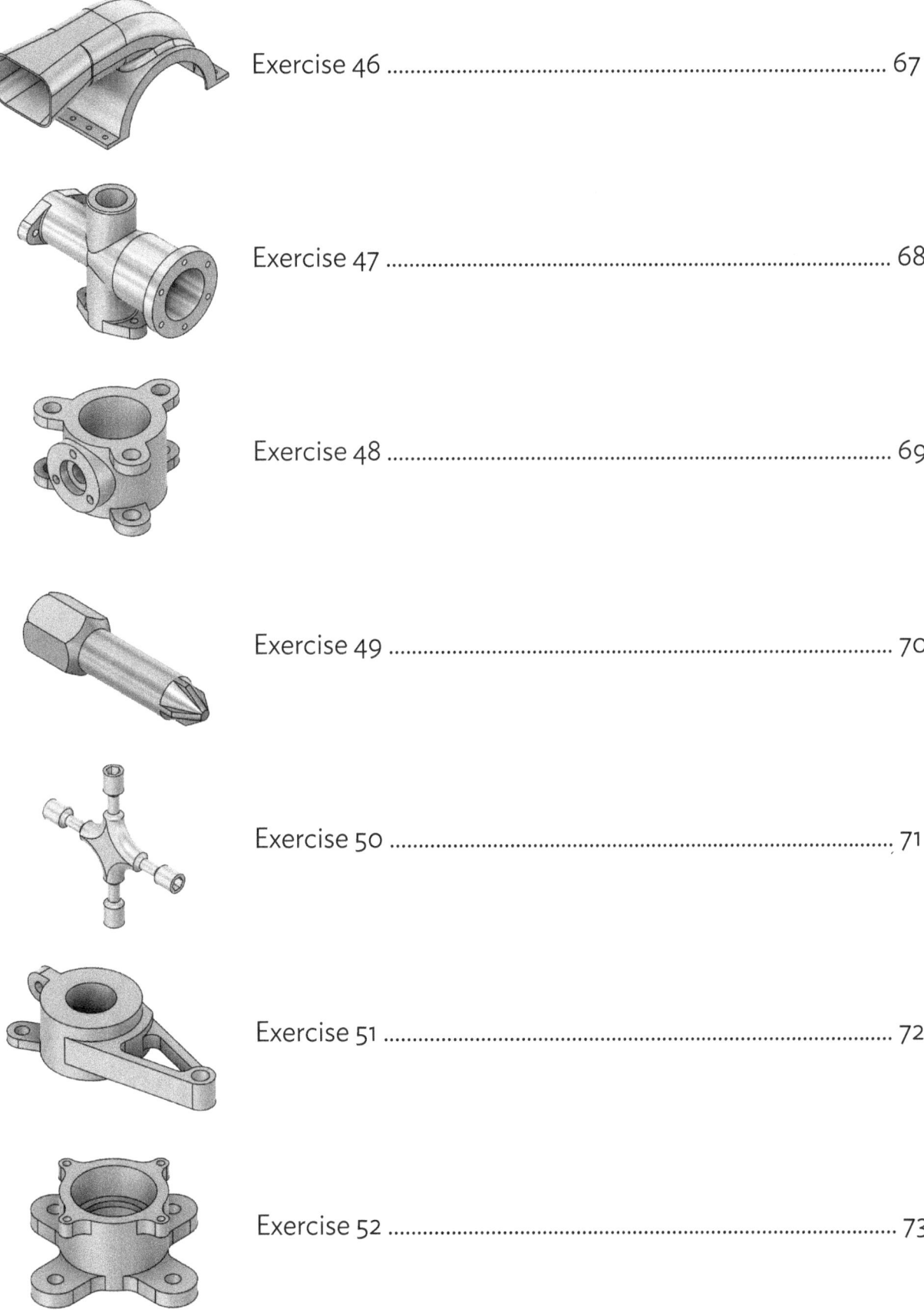

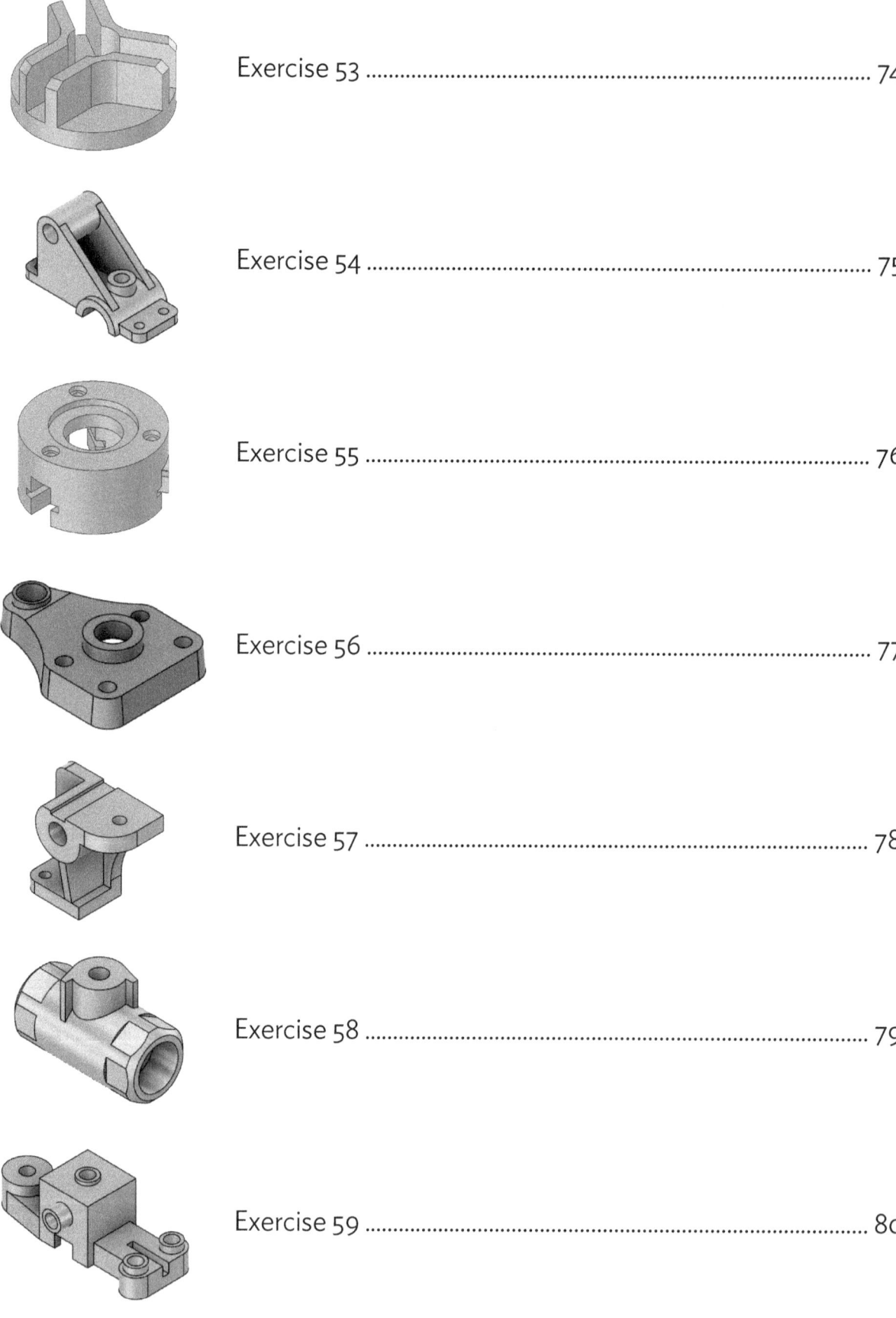

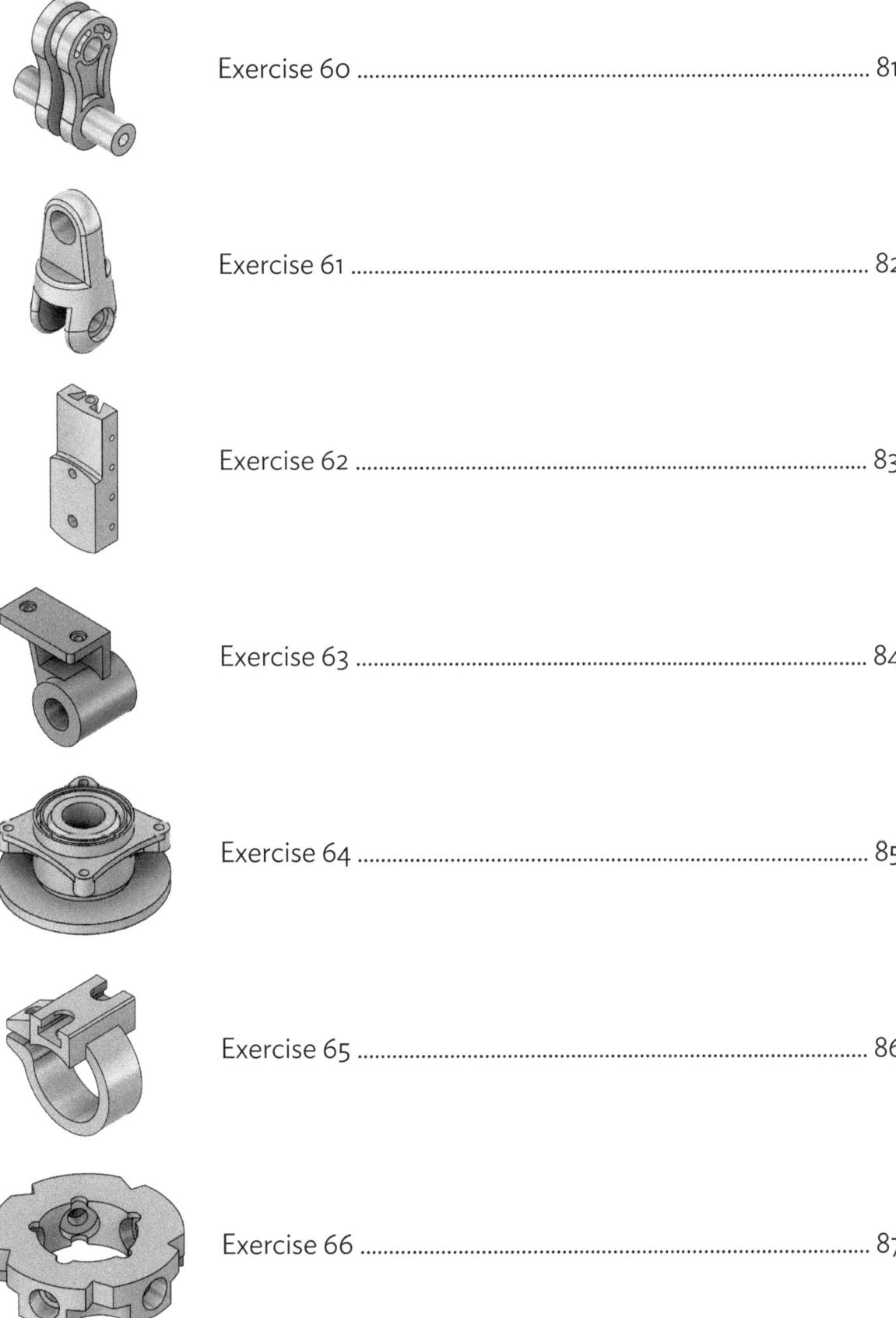

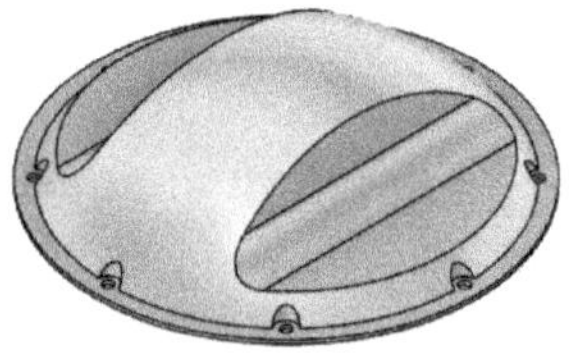

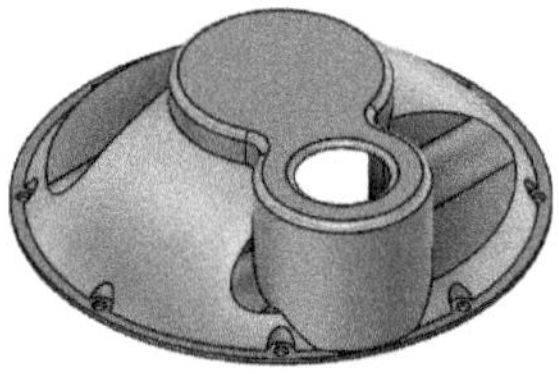

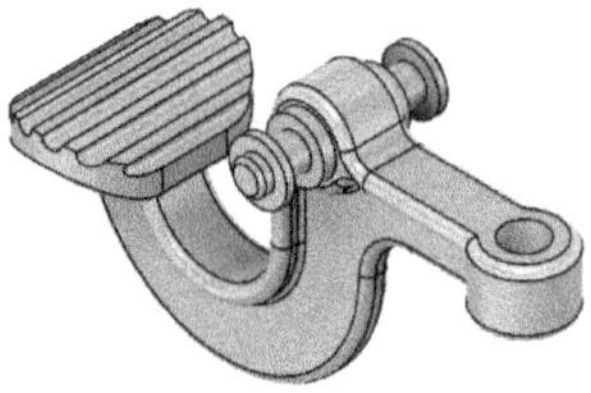

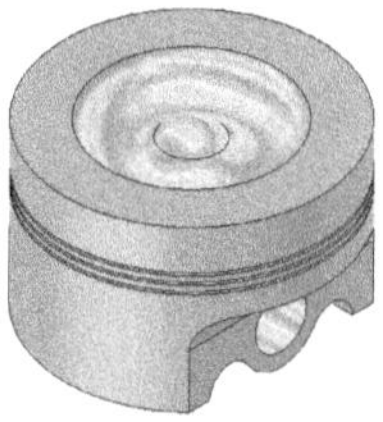

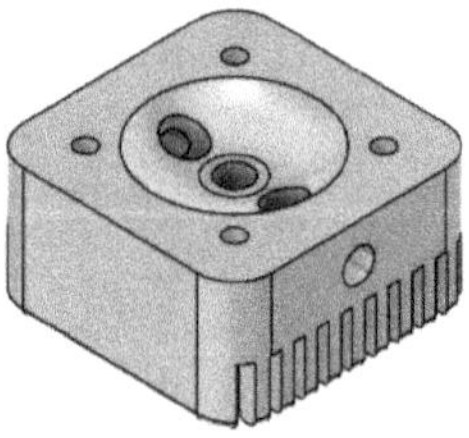

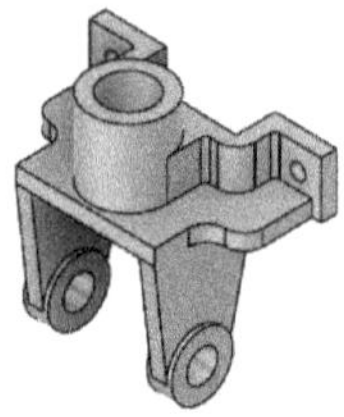

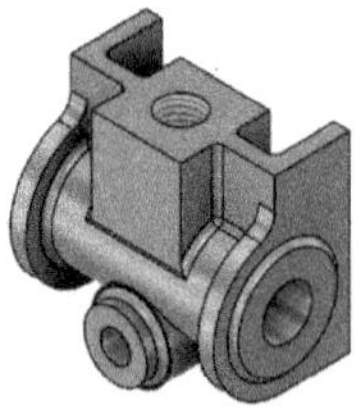

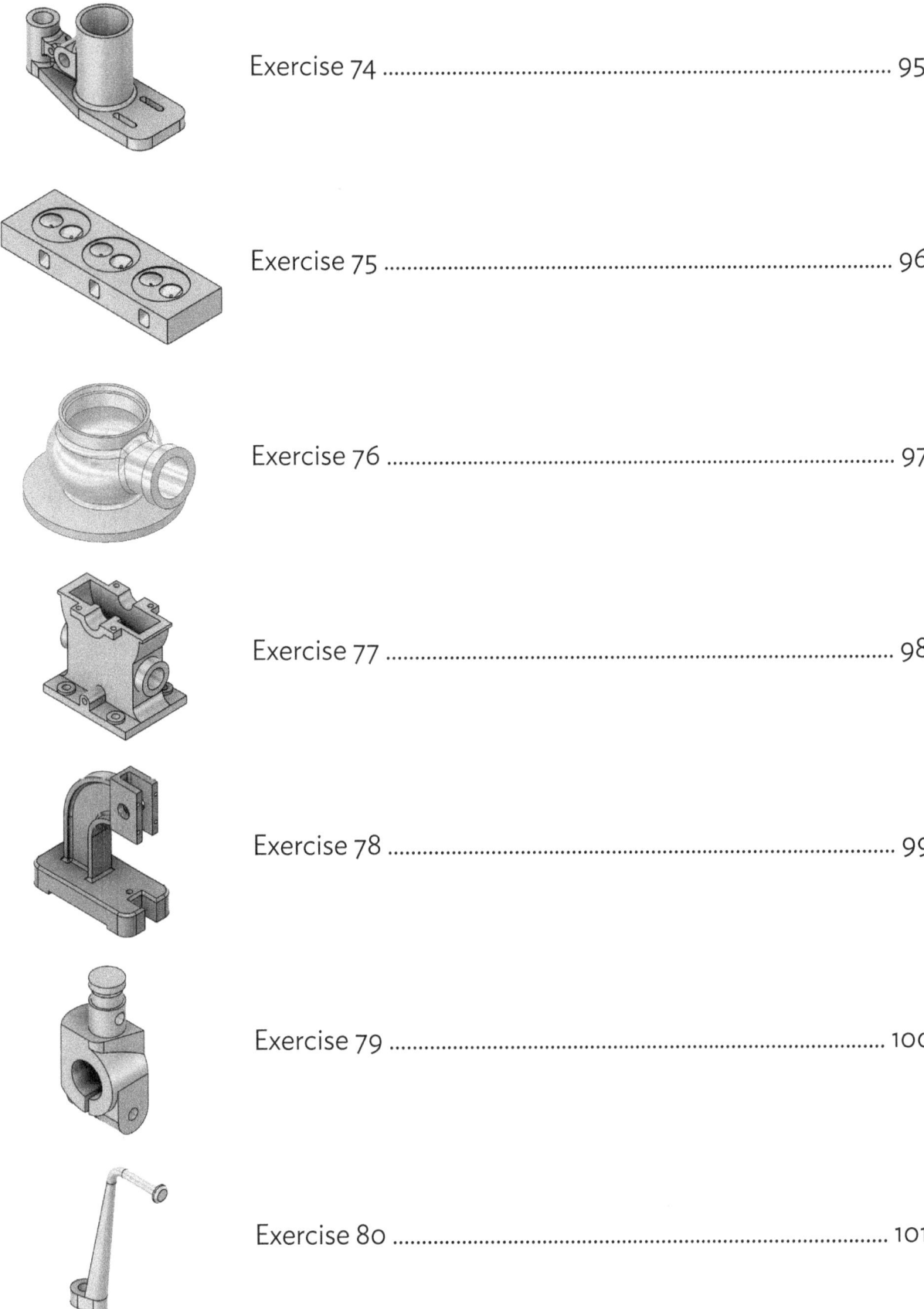

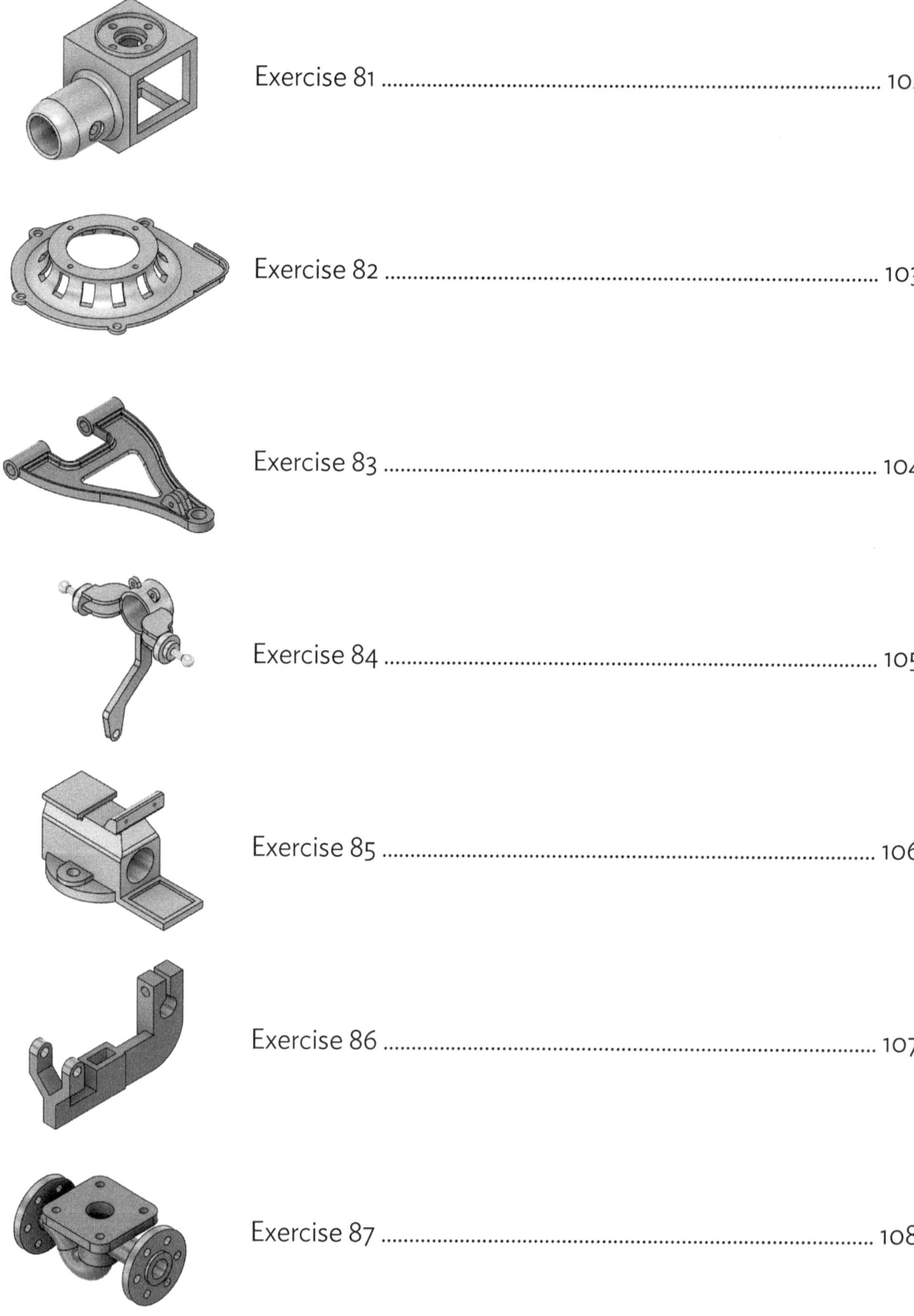

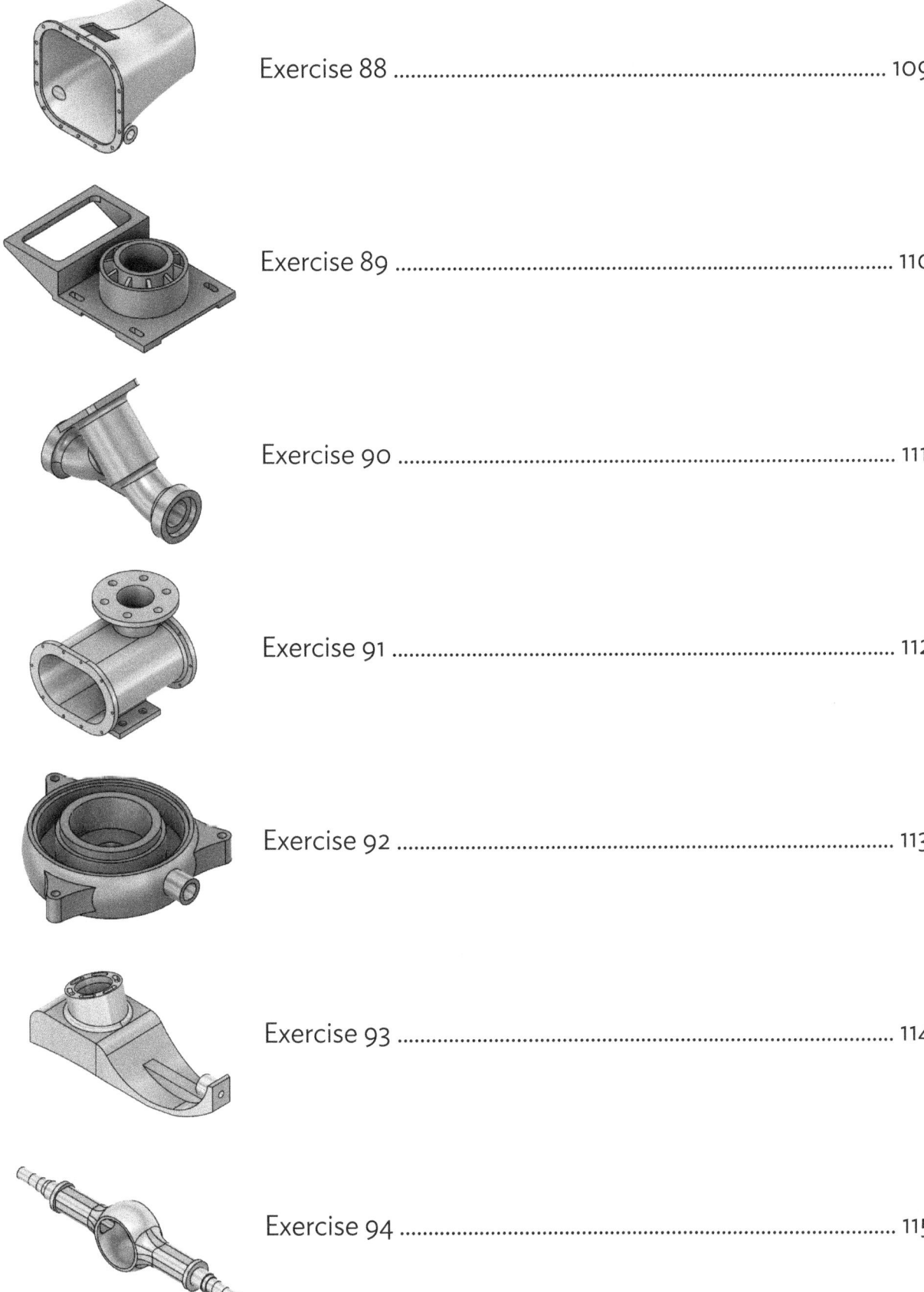

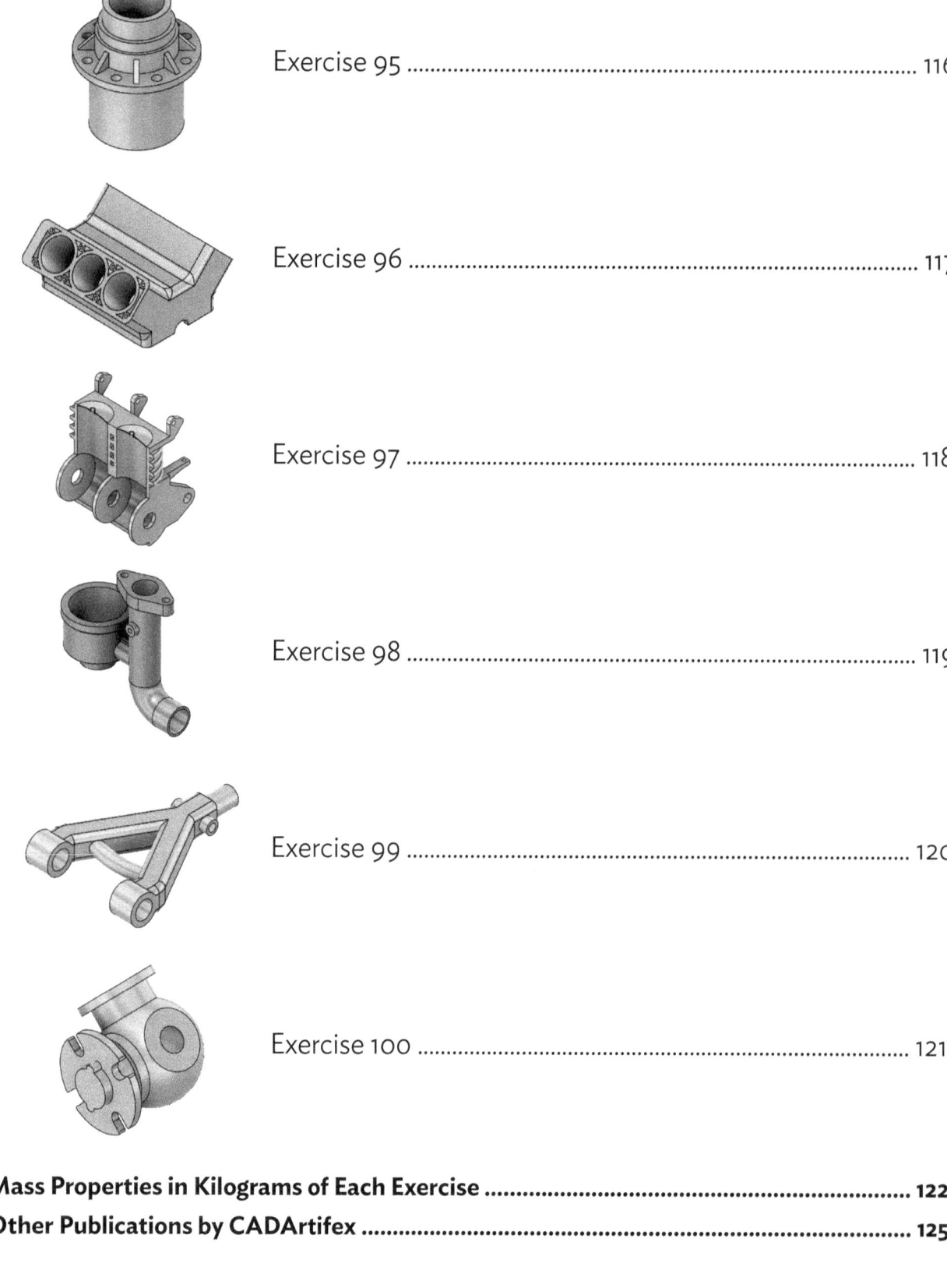

Preface

Autodesk Fusion 360 is a product of Autodesk Inc., one of the biggest providers of technology for the engineering, architecture, construction, manufacturing, media, and entertainment industries. It offers robust software tools for 3D design, engineering, and entertainment industries that let you design, visualize, simulate, and publish your ideas before they are built or created. Moreover, Autodesk continues to develop a comprehensive portfolio of state-of-the-art CAD/CAM/CAE software for the global market.

Autodesk Fusion 360 delivers a rich set of integrated tools that are powerful and intuitive to use. It is the first cloud-based 3D CAD/CAM/CAE software that combines the entire product development cycle into a single cloud-based platform. It allows you to design feature-based, parametric mechanical designs by using simple but highly effective 3D modeling tools. Fusion 360 provides a wide range of tools that allow you to create real-world components and assemblies. These components and assemblies can be converted into 2D engineering drawings for production and used for validating designs by simulating their real-world conditions and assessing the environmental impact of your products. It also enables you to create photorealistic renderings, animations, and toolpaths for CNC machines, in addition to creating rapid prototypes of your design by using the 3D printing workflow.

Autodesk Fusion 360 Exercises - Learn by Practicing (2023-24) book is designed to help engineers and designers interested in learning Autodesk Fusion 360 by practicing 100 real-world mechanical models. This book does not provide step-by-step instructions to design 3D models, instead, it is a practice book that challenges users to first analyze the drawings and then create the models using the powerful toolset of Autodesk Fusion 360. This approach helps users to enhance their design skills and take them to the next level.

Who Should Read This Book

This book is written with a wide range of Autodesk Fusion 360 users in mind, varying from beginners to advanced users. In addition to Autodesk Fusion 360, each exercise of this book can also be designed on any other CAD software such as Autodesk Inventor, CATIA, Creo Parametric, NX, SOLIDWORKS, and Solid Edge.

Prerequisites

To successfully complete the exercises provided in this book, it is essential to possess a solid knowledge of Autodesk Fusion 360. To gain a comprehensive, step-by-step understanding of Autodesk Fusion 360, refer to the '**Autodesk Fusion 360: A Power Guide for Beginners and Intermediate Users (6th Edition)**' textbook published by CADArtifex. A list of all the textbooks published by CADArtifex is given on the final page of this book.

What Is Covered in This Book

Autodesk Fusion 360 Exercises - Learn by Practicing (2023-24) book consists of 100 real-world mechanical models. After creating these models, you will be able to take your design skills to a professional level.

Downloading Exercises

The exercises used in this book are available for free download. To download the exercises, follow the steps given below:

1. Log in to the CADArtifex website (*cadartifex.com/login*) by using your e-mail id and password. If you are a new user, then you first need to register on the CADArtifex website (*cadartifex.com/register*).

2. After logging in to the website, click on **EXERCISES BOOKS** > **Autodesk Fusion 360 Exercises** > **Autodesk Fusion 360 Exercises - Learn by Practicing (2023-24)**. The **Exercises** drop-down list appears for downloading the exercises of this book.

How to Contact the Author

We value your feedback and suggestions. Please email us at *info@cadartifex.com*. You can also log on to our website *www.cadartifex.com* to provide your feedback regarding the book and download the free learning resources.

We thank you for purchasing *Autodesk Fusion 360 Exercises: Learn by Practicing (2023-24)* book and hope that the exercises given in this book will help you to accomplish your professional goals.

Fusion 360
Exercises

Autodesk Fusion 360 Exercises Learn by Practicing (2023-24)

(Design 100 Real-World 3D Models by Practicing)

Each of the 100 exercises in the book can be designed separately. No exercise is a prerequisite for another. The drawing views of the exercises given in this book follow the third angle of projection. You can also download each exercise of the book by login to the CADArtifex website (*cadartifex.com/login*) using your login credentials. After login to the CADArtifex website, click on *EXERCISES BOOKS > Autodesk Fusion 360 Exercises > Autodesk Fusion 360 Exercises - Learn by Practicing (2023-24)*. If you are a new user, then you need to first register on the CADArtifex website (*cadartifex.com/register*).

Exercise 1.

Create the 3D model, as shown in Figure 1. Different views and dimensions are shown in Figure 2. After creating the model, assign the Stainless Steel AISI 304 material and calculate its mass properties. All dimensions are in mm.

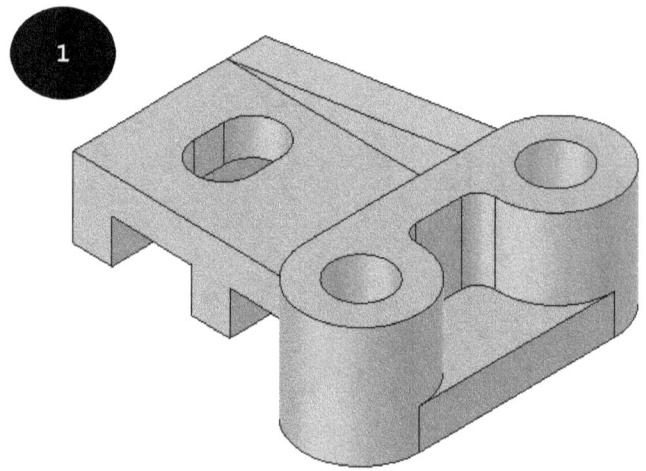

2

MATERIAL: Stainless Steel AISI 304			CADArtifex		
WEIGHT:					
DATE	DRAWN BY:	CHECK BY:	SCALE:	TITLE:	Drawing No.:
			PROJECT:		SHEET: 1 OF 1 / REV.: 0

Exercise 2.

Create the 3D model, as shown in Figure 3. Different views and dimensions are shown in Figure 4. After creating the model, assign the Steel AISI 1020 107 HR material and calculate its mass properties. All dimensions are in mm.

3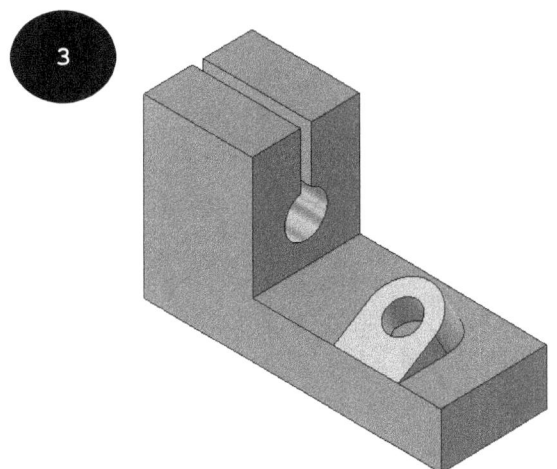

4

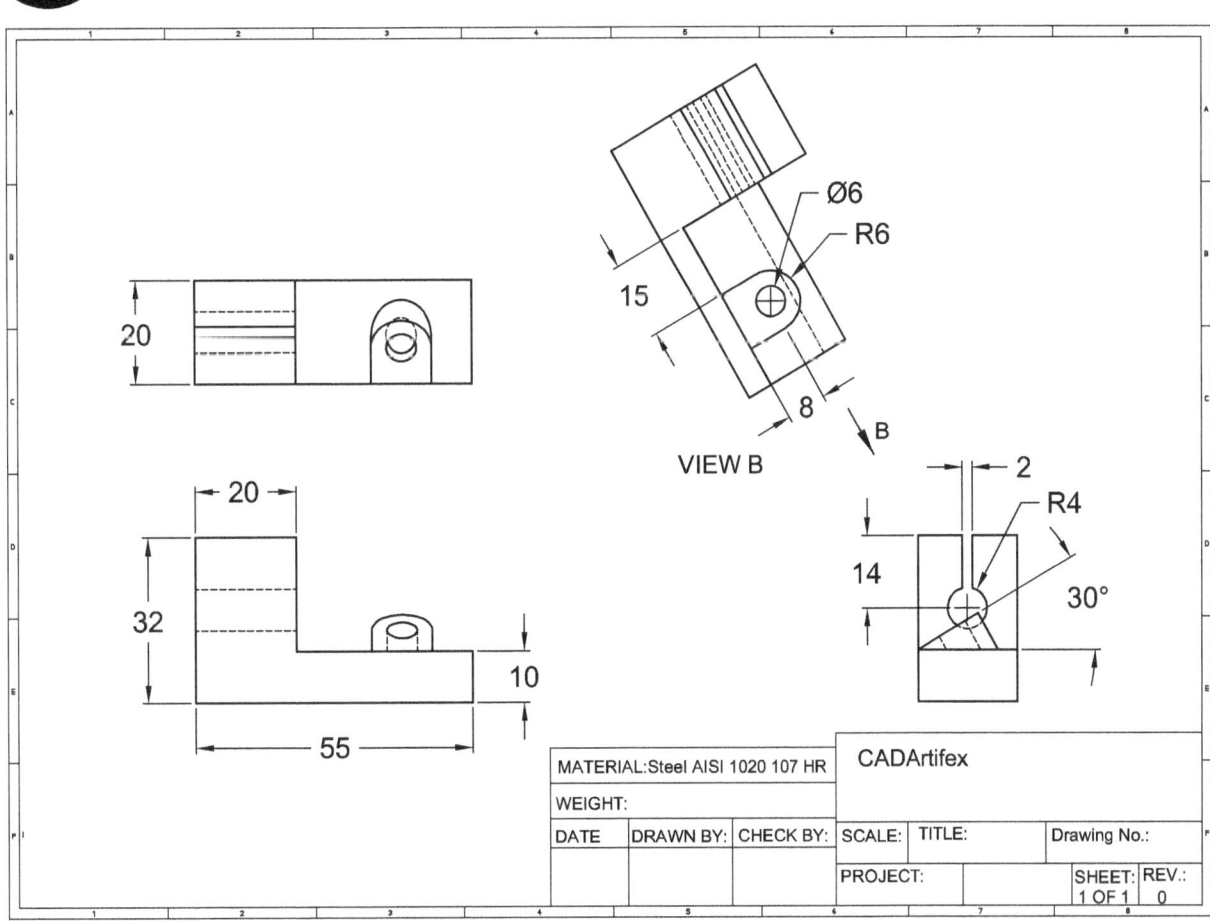

VIEW B

MATERIAL:Steel AISI 1020 107 HR			CADArtifex			
WEIGHT:						
DATE	DRAWN BY:	CHECK BY:	SCALE:	TITLE:		Drawing No.:
			PROJECT:			SHEET: 1 OF 1	REV.: 0

Exercise 3.

Create the 3D model, as shown in Figure 5. Different views of the model and dimensions are shown in Figure 6. After creating the model, assign the Stainless Steel material and calculate its mass properties. All dimensions are in mm.

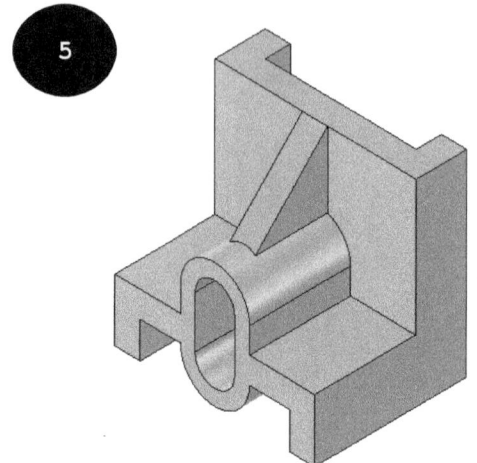

5

6

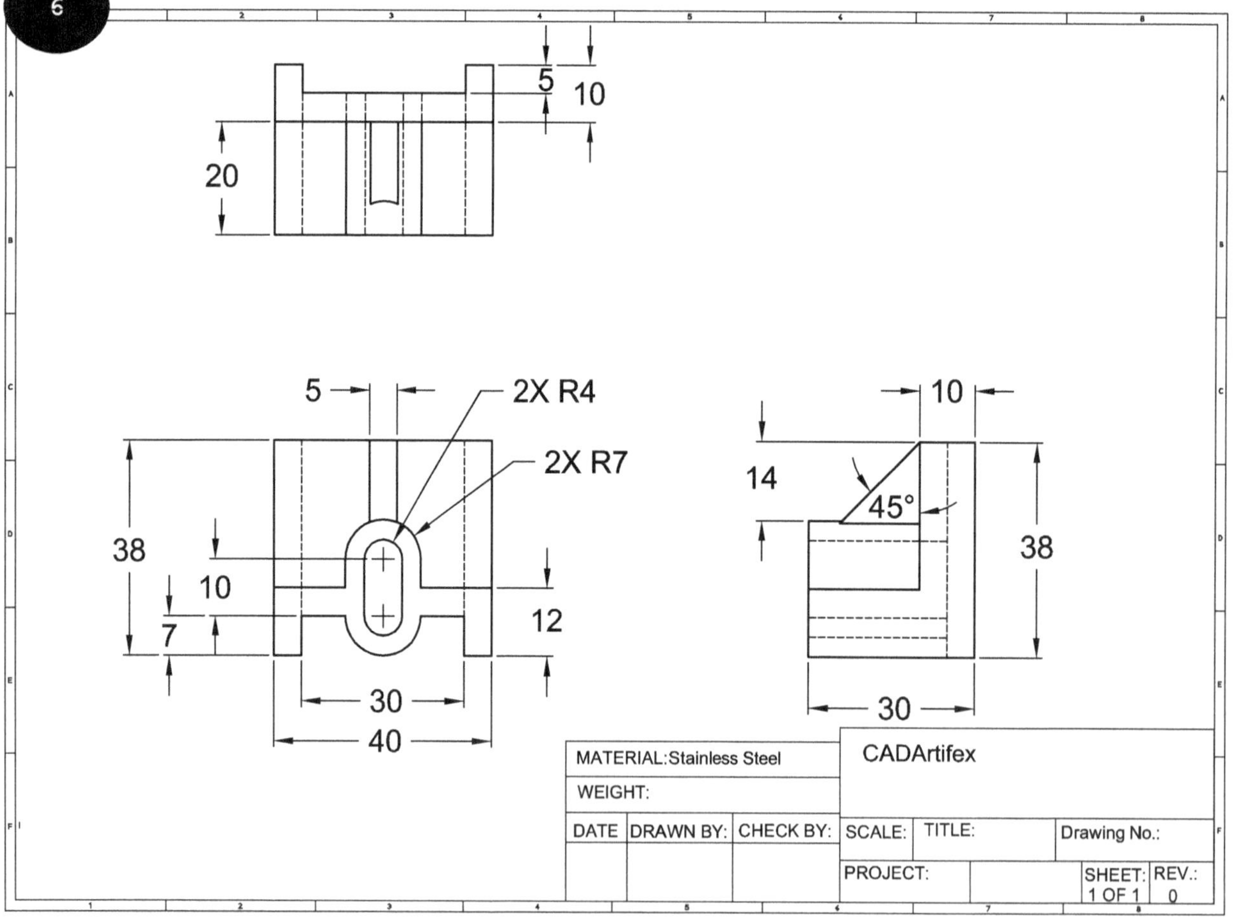

5
10
20

5
2X R4
2X R7
38
10
7
30
40
12

10
14
45°
38
30

MATERIAL:Stainless Steel			CADArtifex		
WEIGHT:					
DATE	DRAWN BY:	CHECK BY:	SCALE:	TITLE:	Drawing No.:
			PROJECT:		SHEET: 1 OF 1	REV.: 0

Exercise 4.

Create the 3D model, as shown in Figure 7. Different views of the model and dimensions are shown in Figure 8. After creating the model, assign the Platinum material and calculate its mass properties. All dimensions are in mm.

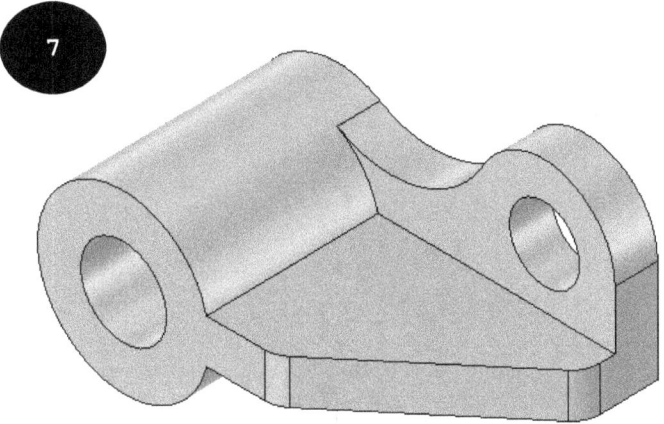

7

8

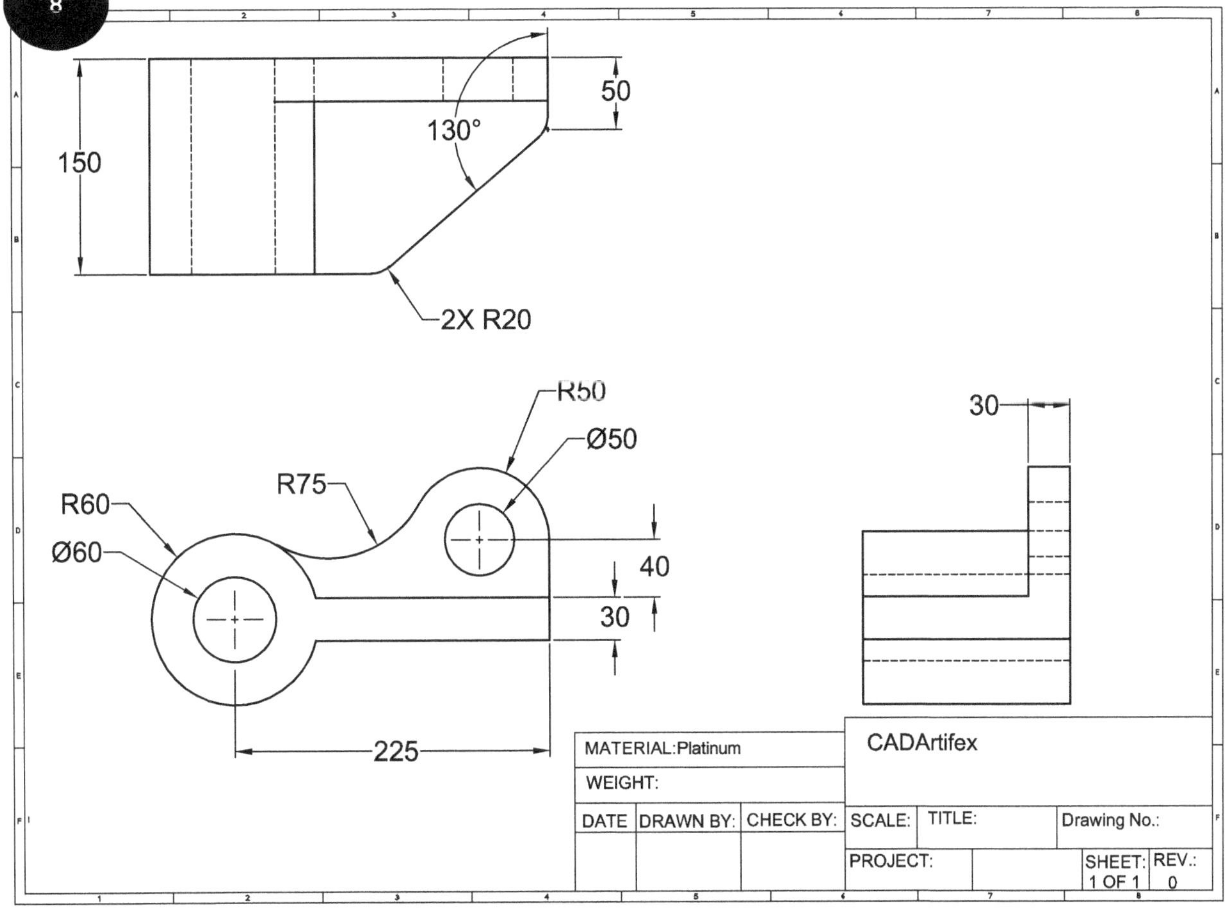

150

130°

50

2X R20

R50

Ø50

R75

R60

Ø60

40

30

30

225

MATERIAL:Platinum			CADArtifex			
WEIGHT:						
DATE	DRAWN BY:	CHECK BY:	SCALE:	TITLE:		Drawing No.:
			PROJECT:			SHEET: REV.: 1 OF 1 0

Exercise 5.

Create the 3D model, as shown in Figure 9. Different views of the model and dimensions are shown in Figure 10. After creating the model, assign the Stainless Steel AISI 202 material and calculate its mass properties. All dimensions are in mm.

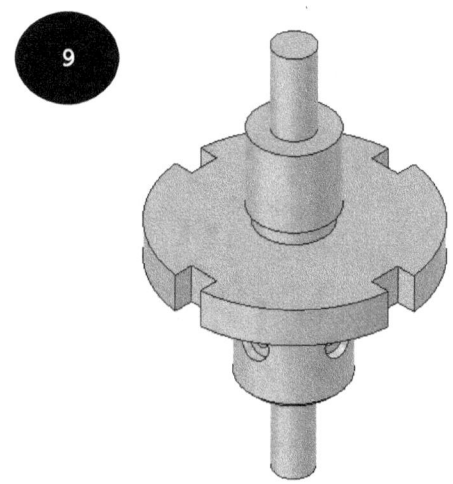

9

10

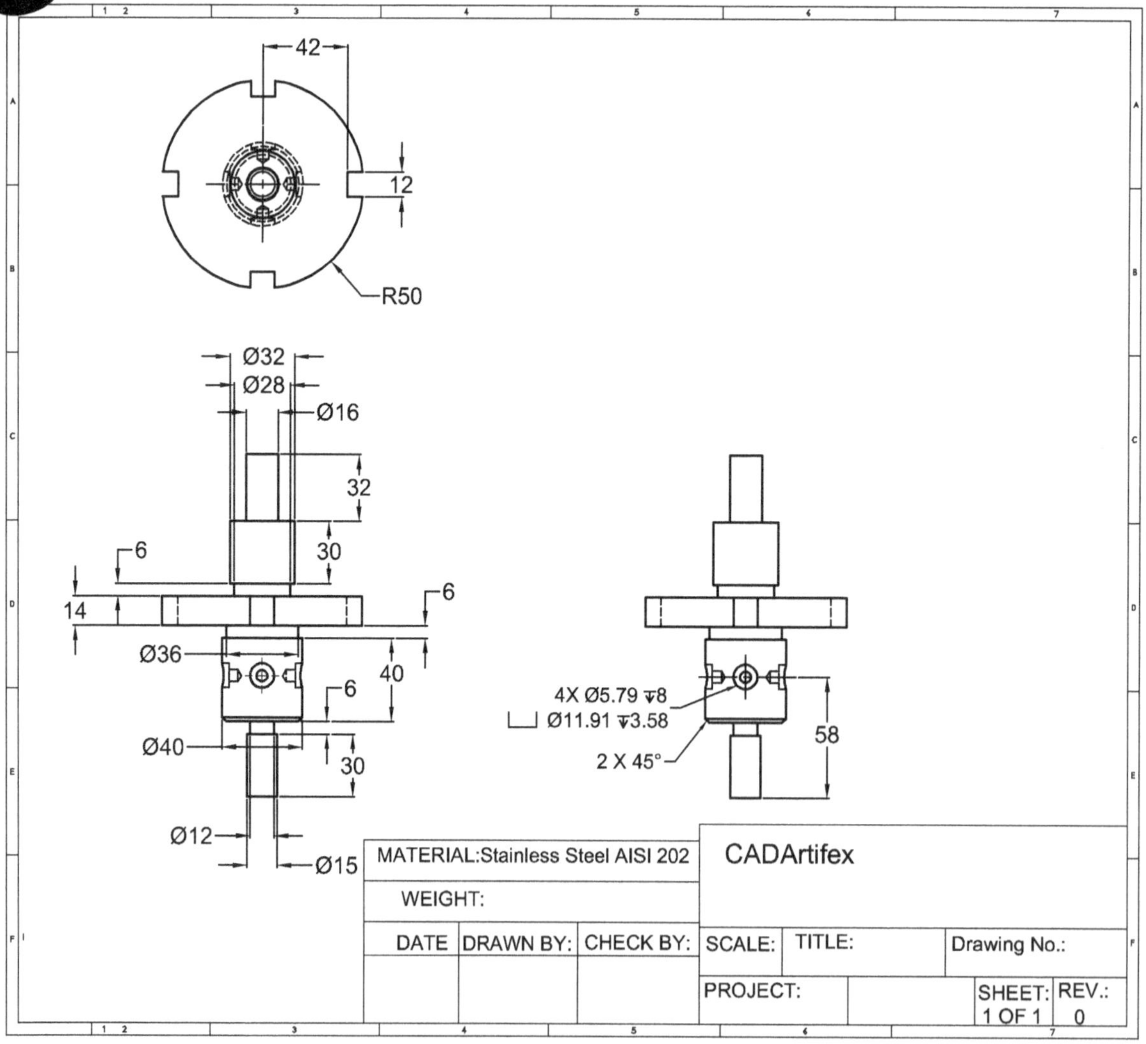

42
12
R50
Ø32
Ø28
Ø16
32
30
6
14
6
Ø36
40
6
Ø40
30
Ø12
Ø15
4X Ø5.79 ↧8
⌴ Ø11.91 ↧3.58
2 X 45°
58
MATERIAL:Stainless Steel AISI 202
WEIGHT:
DATE
DRAWN BY:
CHECK BY:
CADArtifex
SCALE:
TITLE:
Drawing No.:
PROJECT:
SHEET: 1 OF 1
REV.: 0

Exercise 6.

Create the 3D model, as shown in Figure 11. Different views of the model and dimensions are shown in Figure 12. After creating the model, assign the Stainless Steel material and calculate its mass properties. All dimensions are in mm.

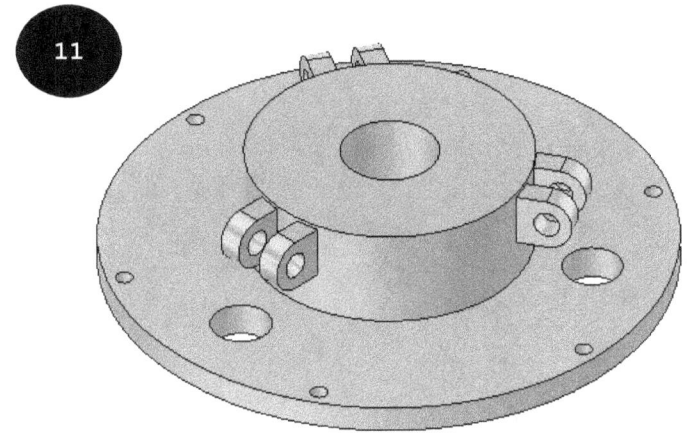

11

12

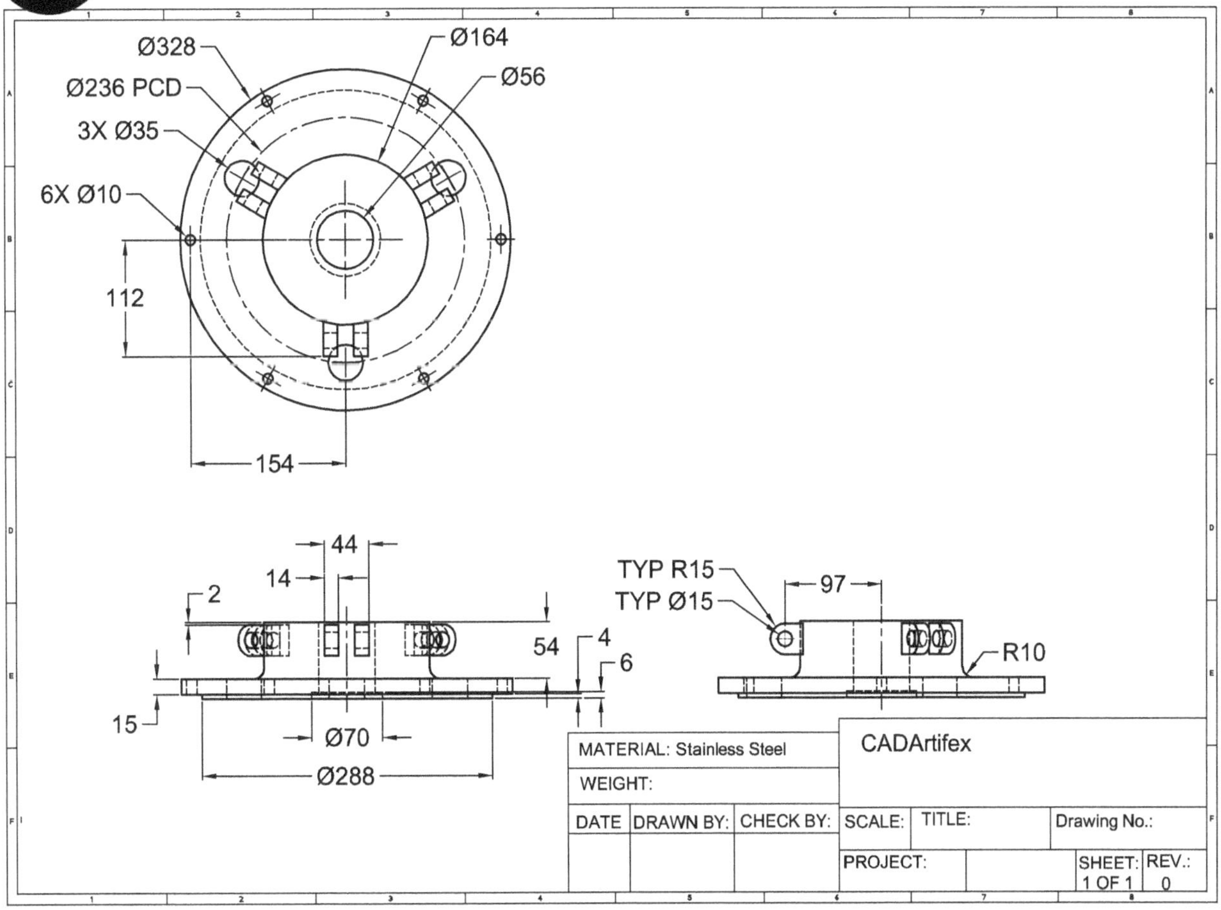

Ø328
Ø164
Ø236 PCD
Ø56
3X Ø35
6X Ø10
112
154

44
14
2
TYP R15
TYP Ø15
97
54
4
6
R10
15
Ø70
Ø288

MATERIAL: Stainless Steel			CADArtifex			
WEIGHT:						
DATE	DRAWN BY:	CHECK BY:	SCALE:	TITLE:		Drawing No.:
			PROJECT:			SHEET: 1 OF 1 / REV.: 0

Exercise 7.

Create the 3D model, as shown in Figure 13. Different views of the model and dimensions are shown in Figure 14. After creating the model, assign the Steel AISI 9262 260 NORM material and calculate its mass properties. All dimensions are in mm.

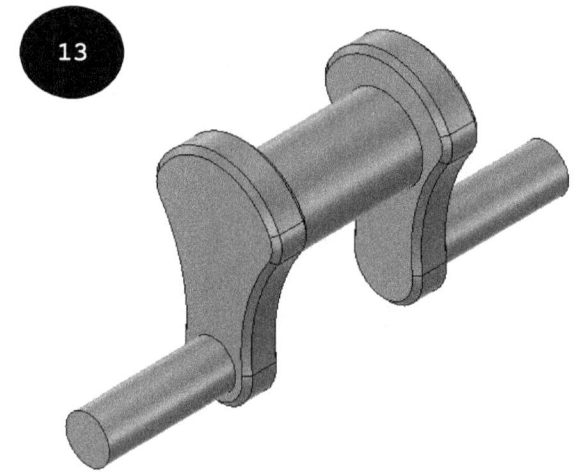

13

14

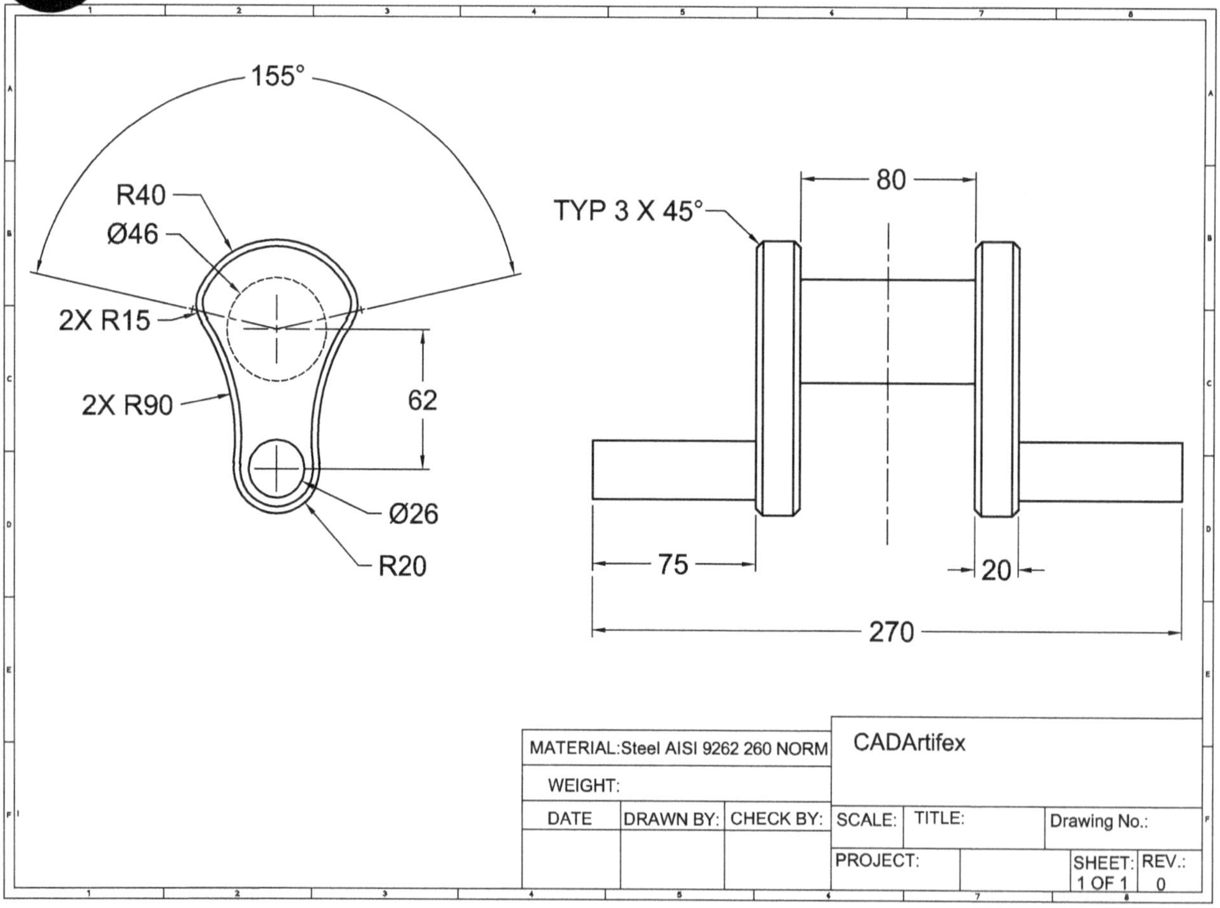

155°

R40
Ø46
2X R15
2X R90
62
Ø26
R20

TYP 3 X 45°
80
75
20
270

MATERIAL:Steel AISI 9262 260 NORM			CADArtifex			
WEIGHT:						
DATE	DRAWN BY:	CHECK BY:	SCALE:	TITLE:		Drawing No.:
			PROJECT:			SHEET: 1 OF 1 / REV.: 0

Exercise 8.

Create the 3D model, as shown in Figure 15. Different views of the model and dimensions are shown in Figure 16. After creating the model, assign the Carbon Steel material and calculate its mass properties. All dimensions are in mm.

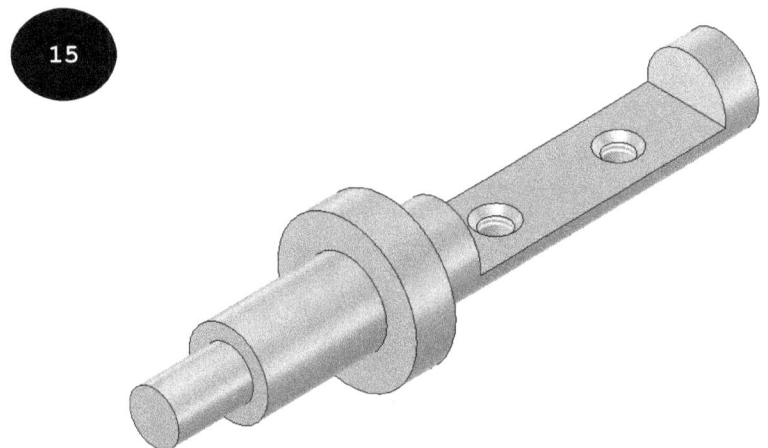

15

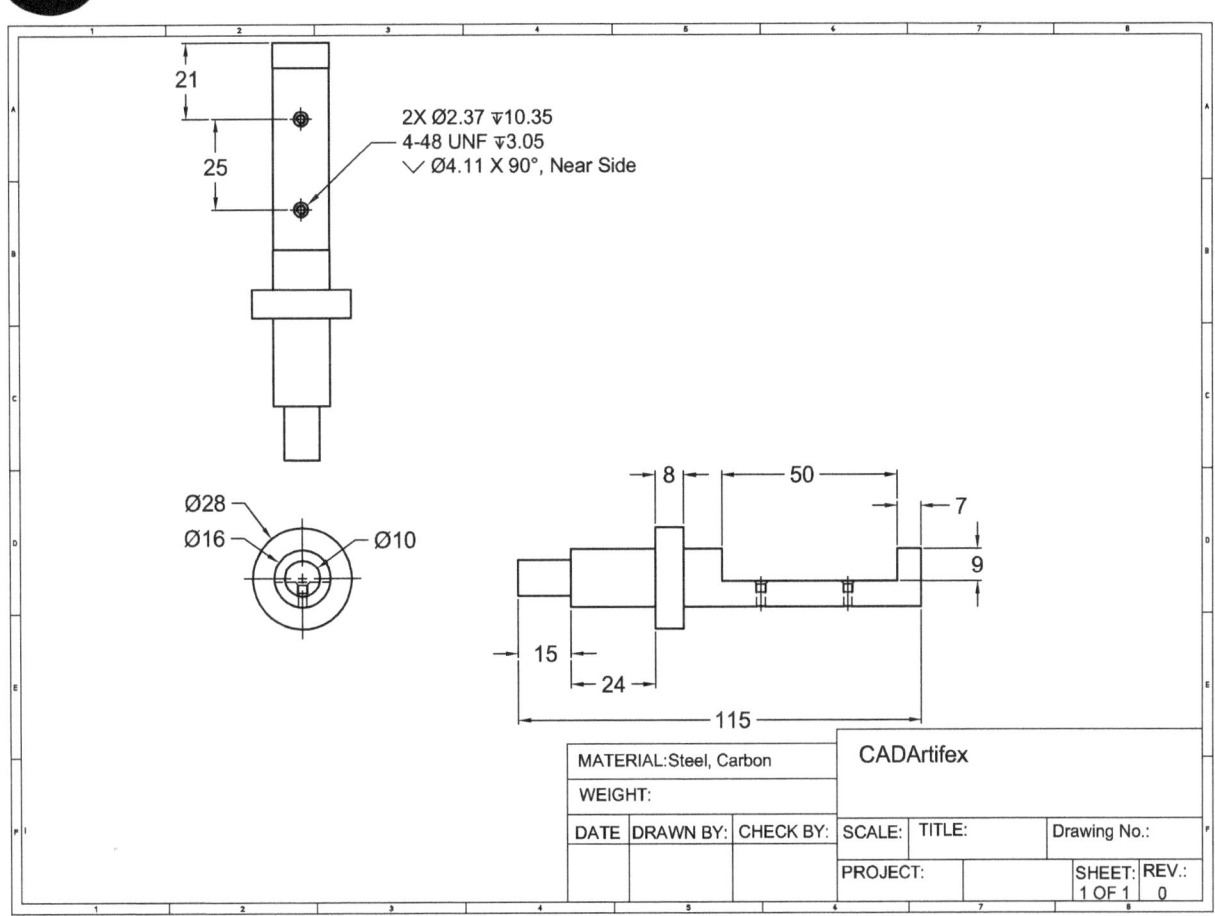

16

Exercise 9.

Create the 3D model, as shown in Figure 17. Different views of the model and dimensions are shown in Figure 18. After creating the model, assign the Stainless Steel AISI 304 material and calculate its mass properties. All dimensions are in mm.

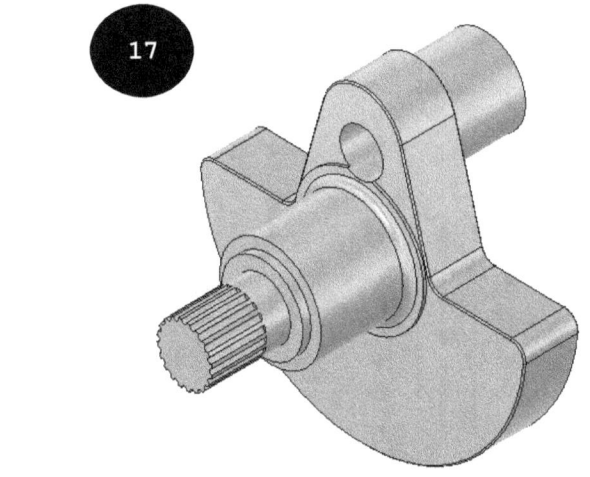

17

18

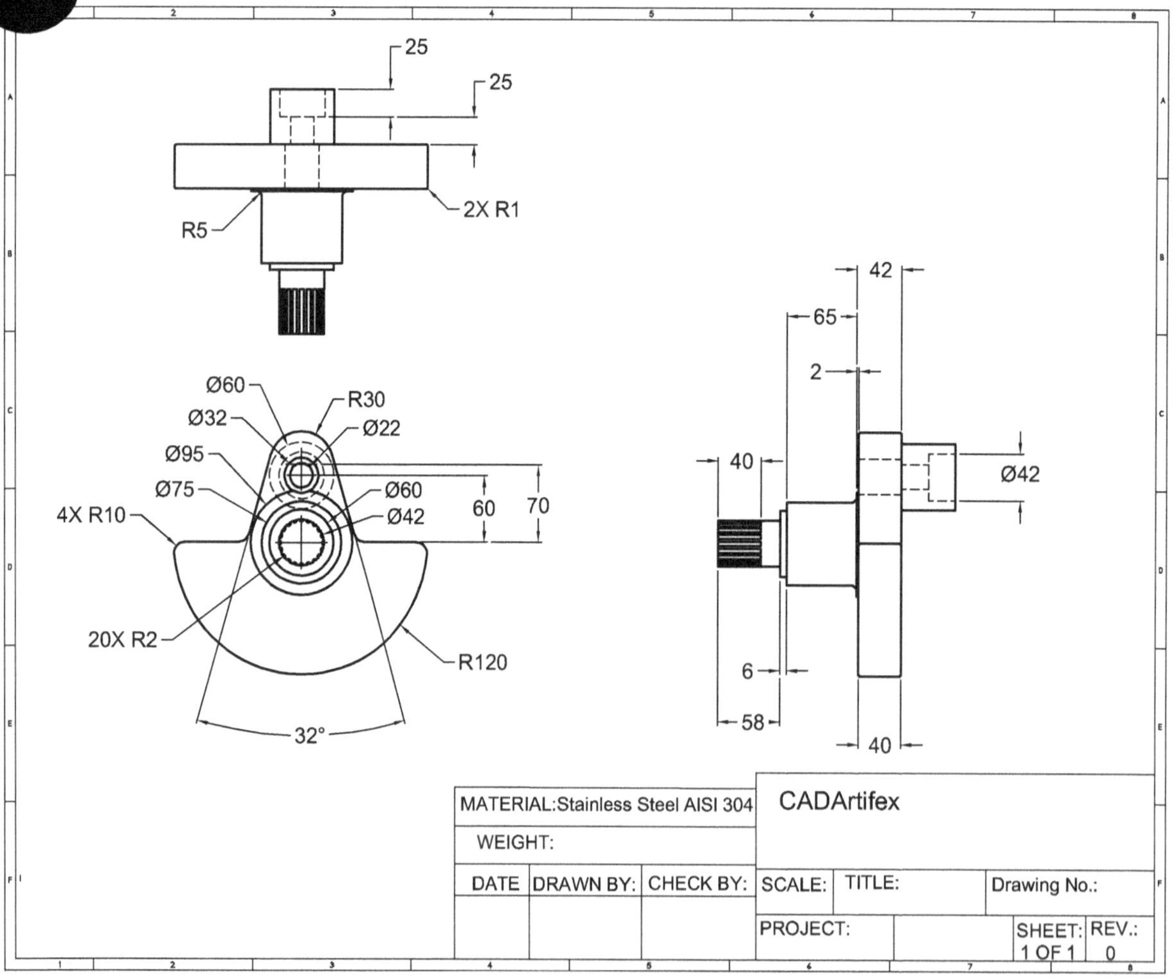

25
25
2X R1
R5
42
65
2
Ø60
Ø32
R30
Ø22
Ø95
Ø75
Ø60
Ø42
4X R10
60
70
40
Ø42
20X R2
R120
6
58
32°
40
MATERIAL:Stainless Steel AISI 304
CADArtifex
WEIGHT:
DATE
DRAWN BY:
CHECK BY:
SCALE:
TITLE:
Drawing No.:
PROJECT:
SHEET: 1 OF 1
REV.: 0

Exercise 10.

Create the 3D model, as shown in Figure 19. Different views of the model and dimensions are shown in Figure 20. After creating the model, assign the Steel AISI 1020 107 HR material and calculate its mass properties. All dimensions are in mm.

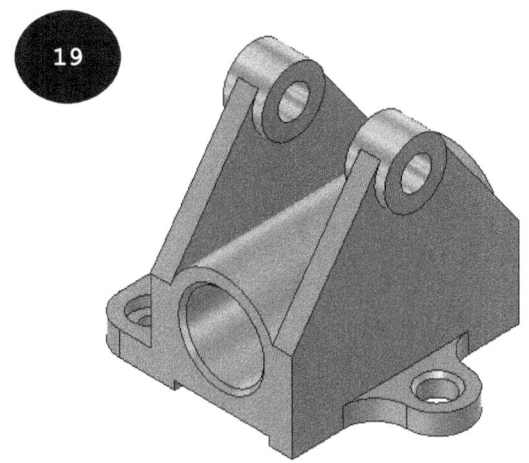

19

20

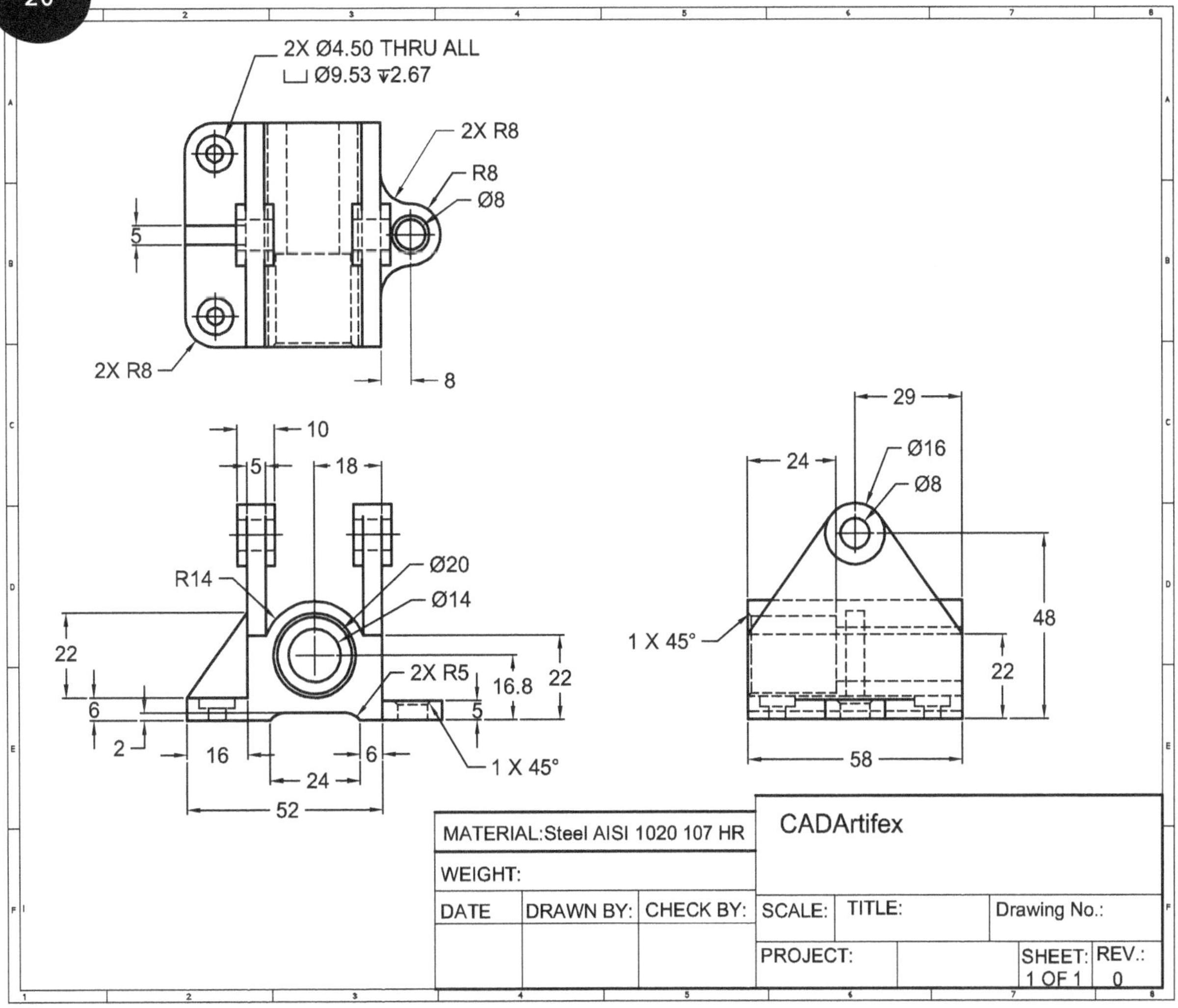

2X Ø4.50 THRU ALL
⌴ Ø9.53 ▾2.67

2X R8

R8
Ø8

5

2X R8

8

10

5
18

R14
Ø20
Ø14

22

2X R5
16.8 22

6
5

2
16
6

24

1 X 45°

52

29

Ø16
Ø8

24

1 X 45°

48

22

58

MATERIAL: Steel AISI 1020 107 HR			CADArtifex		
WEIGHT:					
DATE	DRAWN BY:	CHECK BY:	SCALE:	TITLE:	Drawing No.:
			PROJECT:		SHEET: 1 OF 1	REV.: 0

Exercise 11.

Create the 3D model, as shown in Figure 21. Different views of the model and dimensions are shown in Figure 22. After creating the model, assign the Steel, Alloy material and calculate its mass properties. All dimensions are in mm.

21

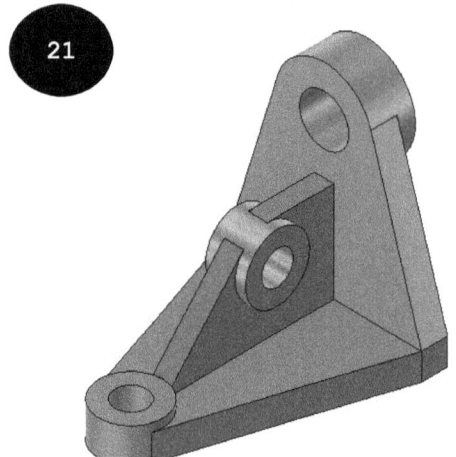

22

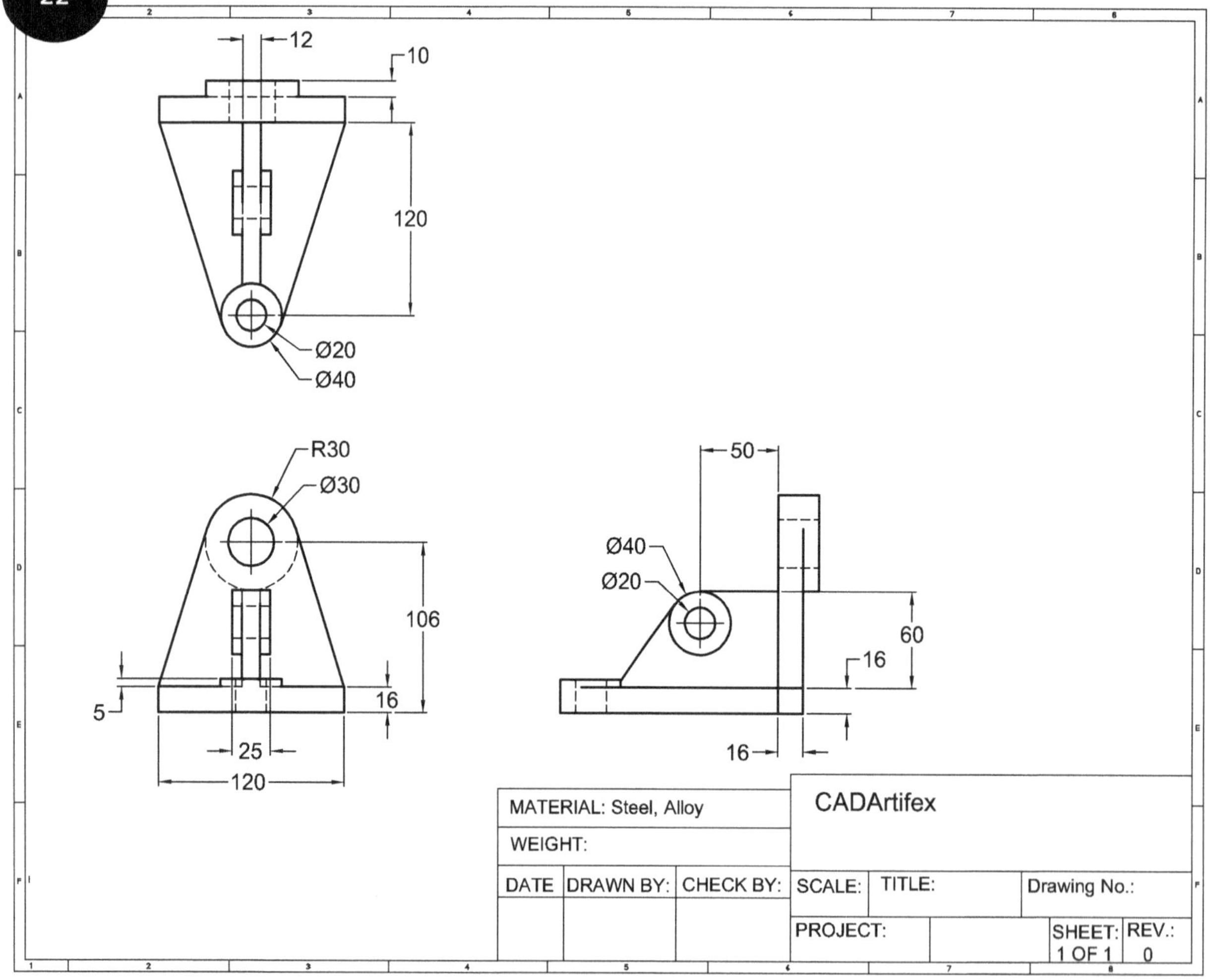

R30
Ø30

12
10
120
Ø20
Ø40

50
Ø40
Ø20
60
16
16

106
5
16
25
120

MATERIAL: Steel, Alloy			CADArtifex			
WEIGHT:						
DATE	DRAWN BY:	CHECK BY:	SCALE:	TITLE:		Drawing No.:
			PROJECT:			SHEET: 1 OF 1 / REV.: 0

Exercise 12.

Create the 3D model, as shown in Figure 23. Different views of the model and dimensions are shown in Figure 24. After creating the model, assign the Steel, Carbon material and calculate its mass properties. All dimensions are in mm.

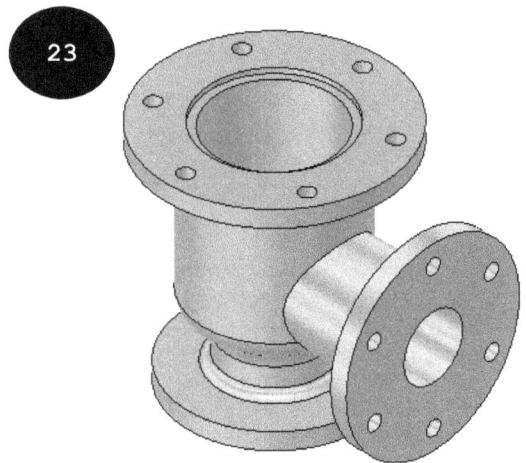

23

24

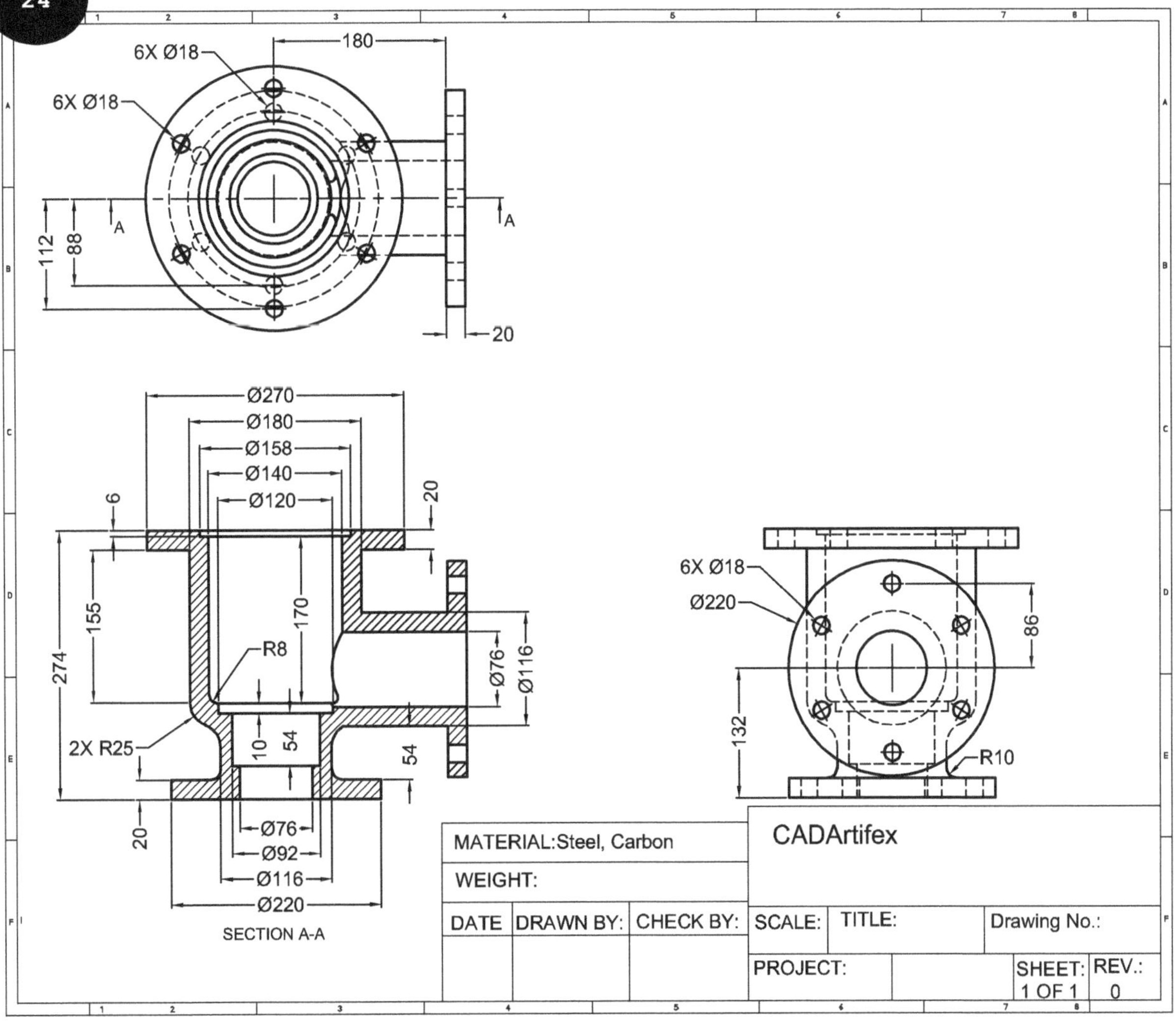

SECTION A-A

MATERIAL:Steel, Carbon			CADArtifex			
WEIGHT:						
DATE	DRAWN BY:	CHECK BY:	SCALE:	TITLE:		Drawing No.:
			PROJECT:			SHEET: 1 OF 1	REV.: 0

6X Ø18
6X Ø18
180
112
88
A
A
20

Ø270
Ø180
Ø158
Ø140
Ø120
6
20
155
274
170
R8
2X R25
10
54
Ø76
Ø116
54
20
Ø76
Ø92
Ø116
Ø220

6X Ø18
Ø220
86
132
R10

Exercise 13.

Create the 3D model, as shown in Figure 25. Different views of the model and dimensions are shown in Figure 26. After creating the model, assign the Steel material and calculate its mass properties. All dimensions are in mm.

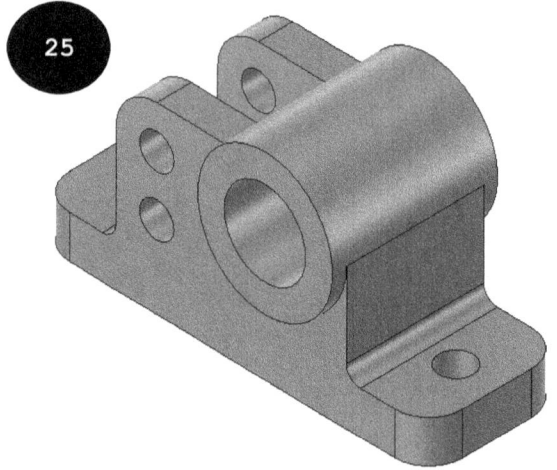

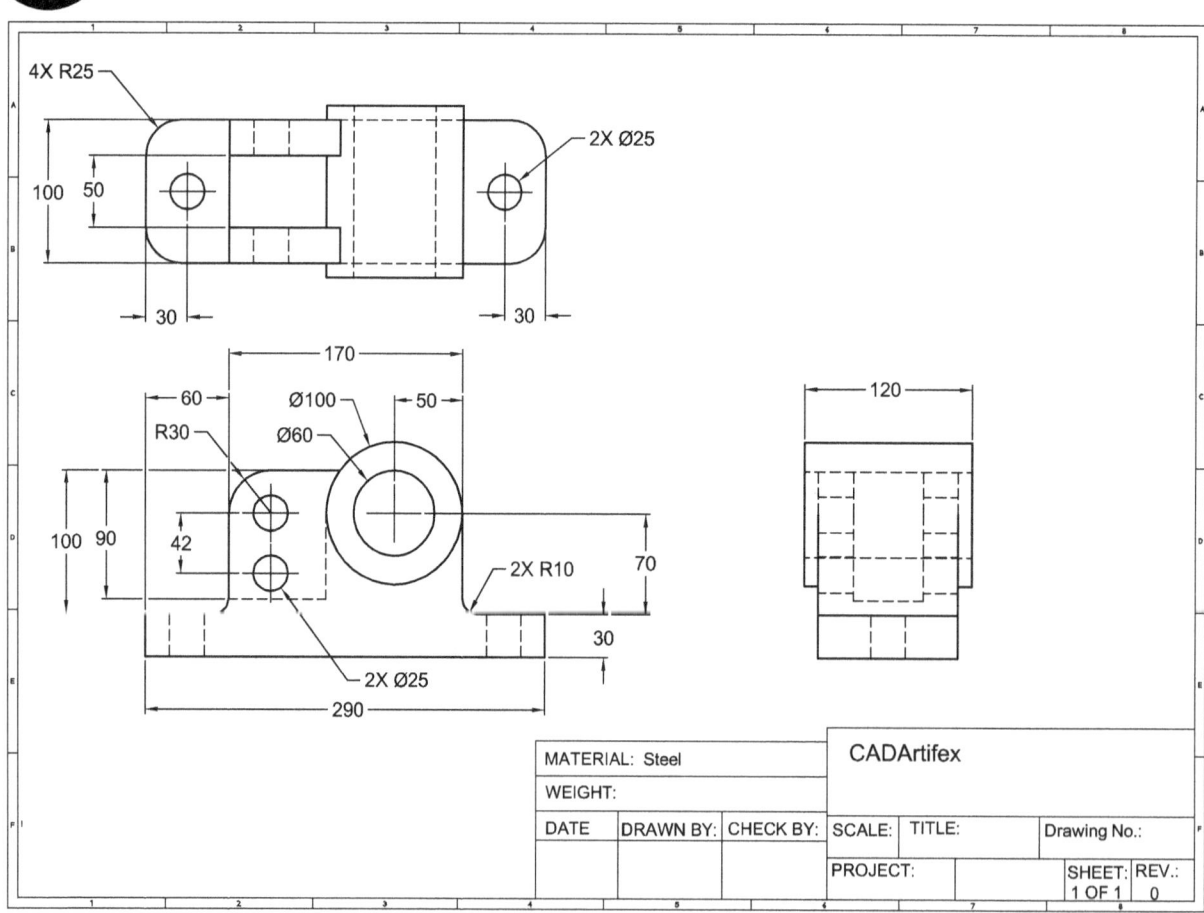

Exercise 14.

Create the 3D model, as shown in Figure 27. Different views of the model and dimensions are shown in Figure 28. After creating the model, assign the Steel material and calculate its mass properties. All dimensions are in mm.

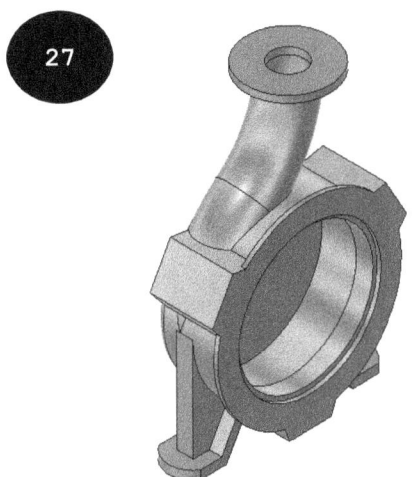

27

28

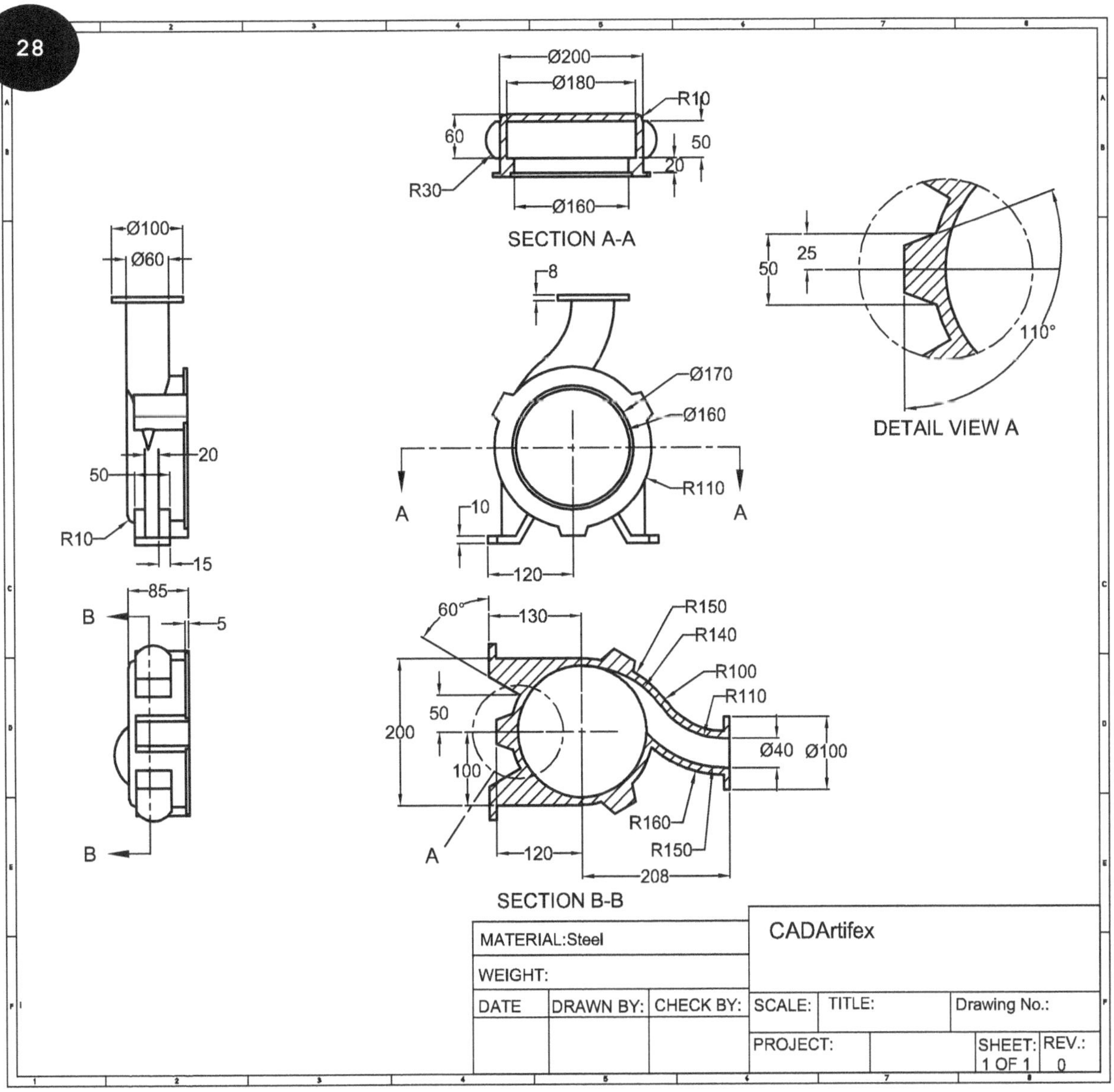

Ø200
Ø180
R10
60
50
R30
20
Ø160

SECTION A-A

Ø100
Ø60

8

Ø170
Ø160

20
50
R110
R10
A   10   A
15   120

85
B
5

50
25
110°

DETAIL VIEW A

B

60°   130   R150
R140
50   R100
200   R110
100   Ø40 Ø100
R160
A   120   R150
208

SECTION B-B

MATERIAL:Steel			CADArtifex		
WEIGHT:					
DATE	DRAWN BY:	CHECK BY:	SCALE:	TITLE:	Drawing No.:
			PROJECT:		SHEET: 1 OF 1	REV.: 0

Exercise 15.

Create the 3D model, as shown in Figure 29. Different views of the model and dimensions are shown in Figure 30. After creating the model, assign the Steel, Alloy material and calculate its mass properties. All dimensions are in mm.

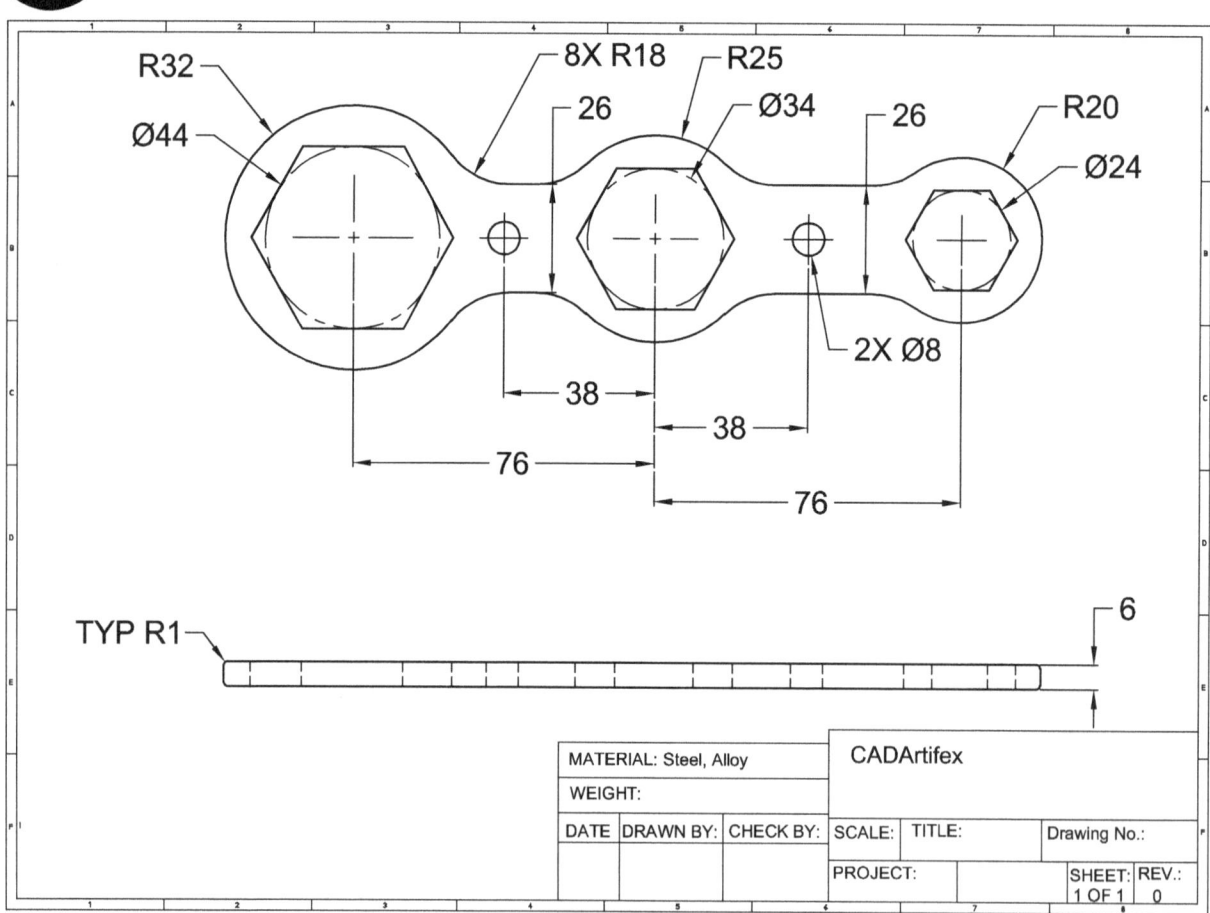

Exercise 16.

Create the 3D model, as shown in Figure 31. Different views of the model and dimensions are shown in Figure 32. After creating the model, assign the Steel, Carbon material and calculate its mass properties. All dimensions are in mm.

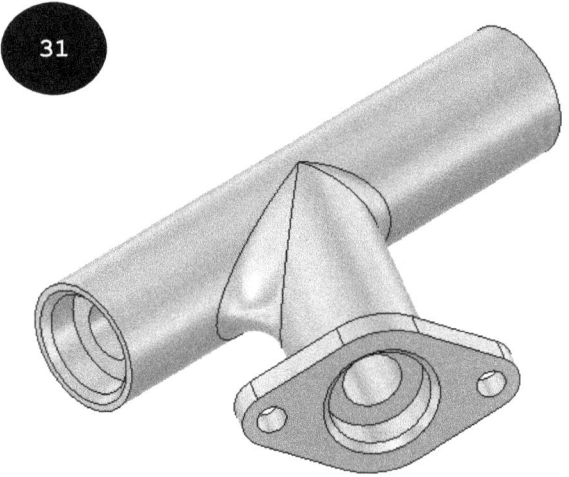

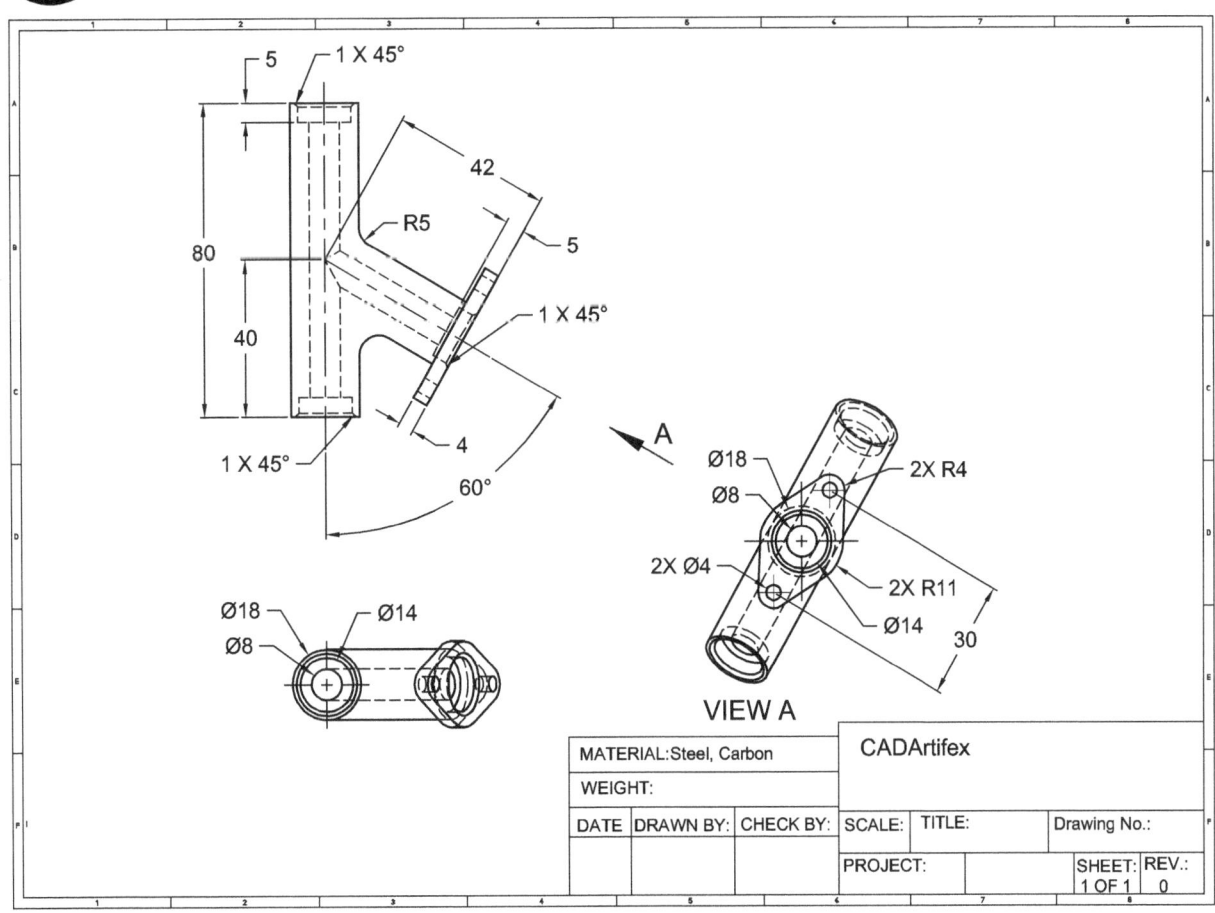

Exercise 17.

Create the 3D model, as shown in Figure 33. Different views of the model and dimensions are shown in Figure 34. After creating the model, assign the Stainless Steel AISI 304 material and calculate its mass properties. All dimensions are in mm.

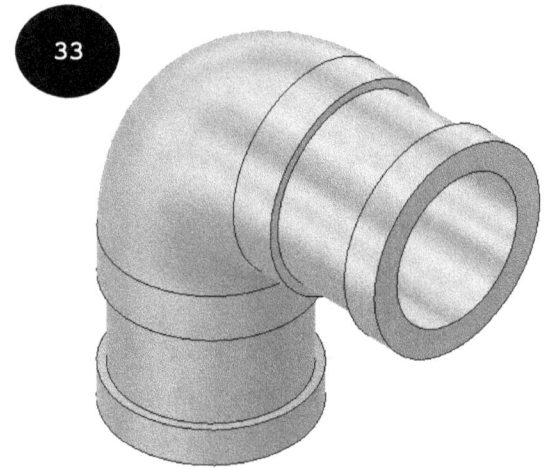

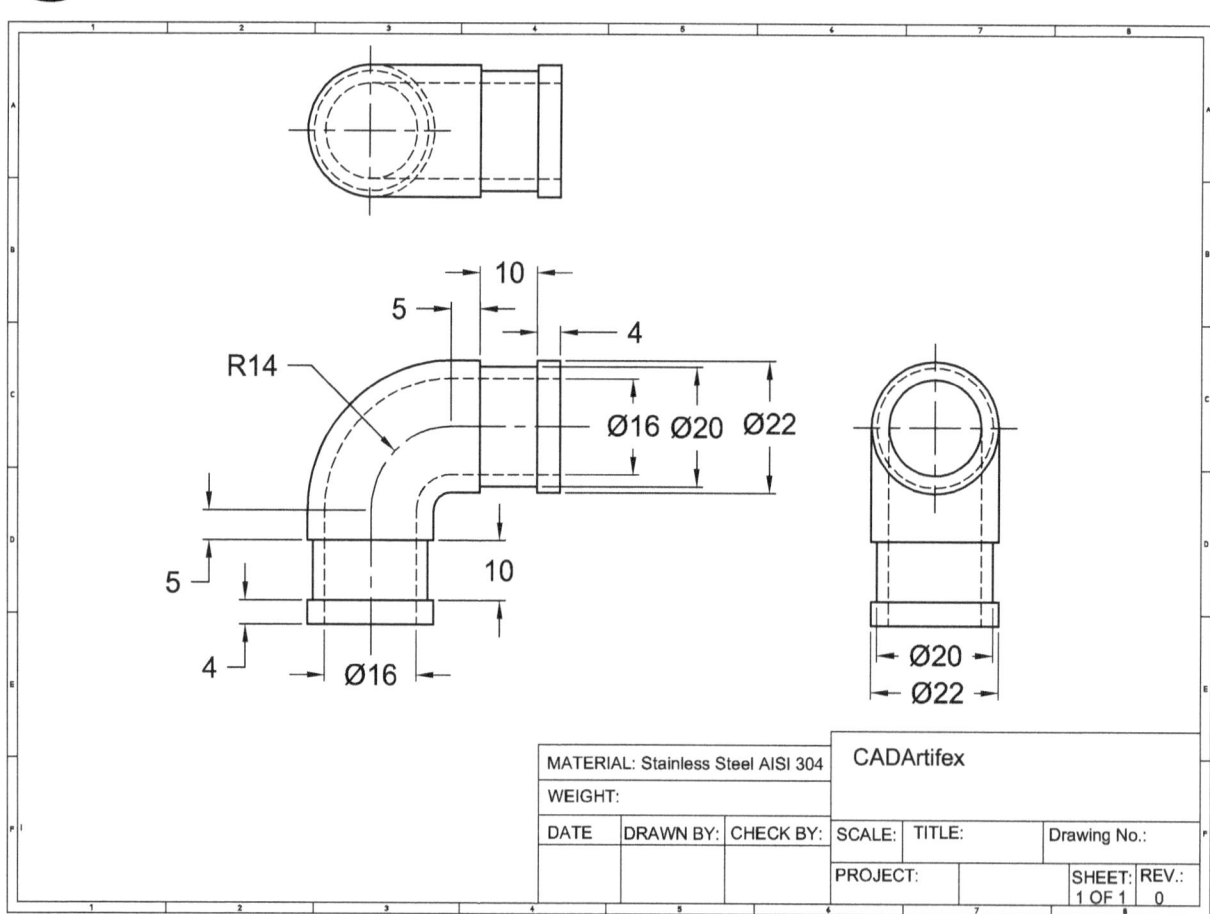

Exercise 18.

Create the 3D model, as shown in Figure 35. Different views of the model and dimensions are shown in Figure 36. After creating the model, assign the Steel AISI 1020 107 HR material and calculate its mass properties. All dimensions are in mm.

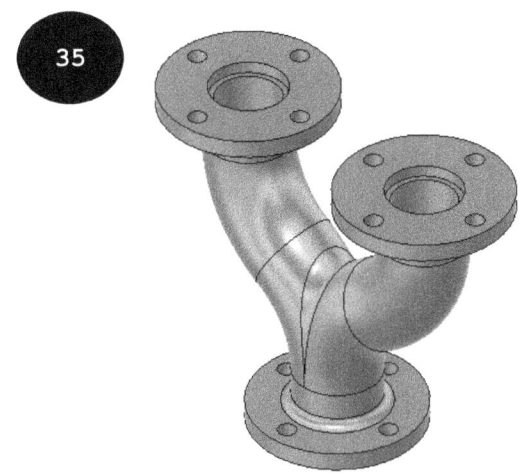

35

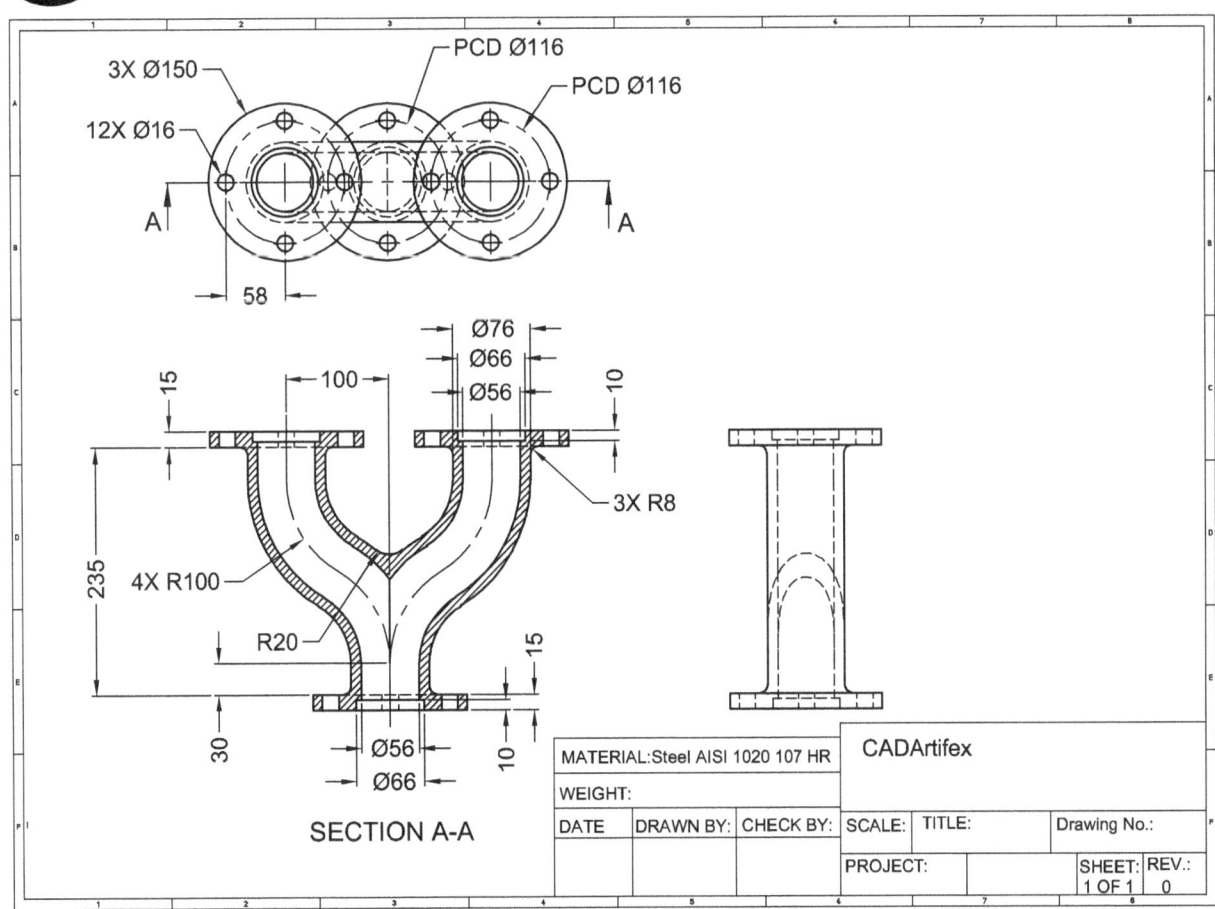

36

SECTION A-A

Exercise 19.

Create the 3D model, as shown in Figure 37. Different views of the model and dimensions are shown in Figure 38. After creating the model, assign the Steel, Alloy material and calculate its mass properties. All dimensions are in mm.

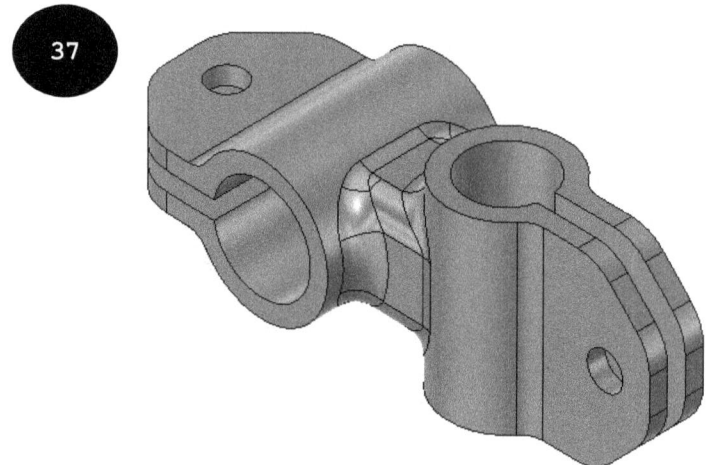

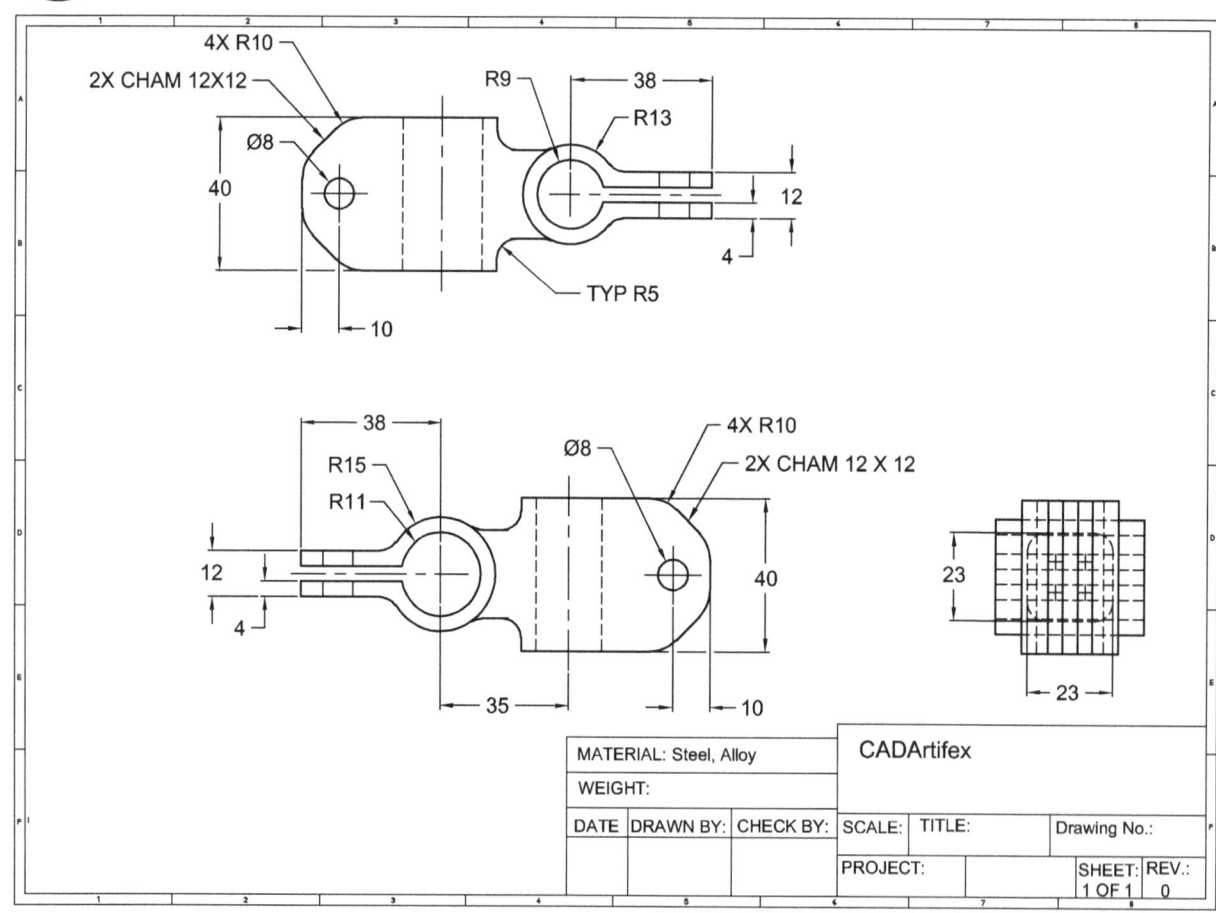

Exercise 20.

Create the 3D model, as shown in Figure 39. Different views of the model and dimensions are shown in Figure 40. After creating the model, assign the Steel, Carbon material and calculate its mass properties. All dimensions are in mm.

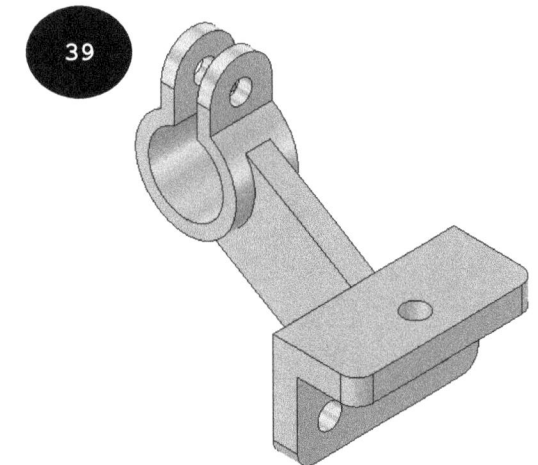

39

40

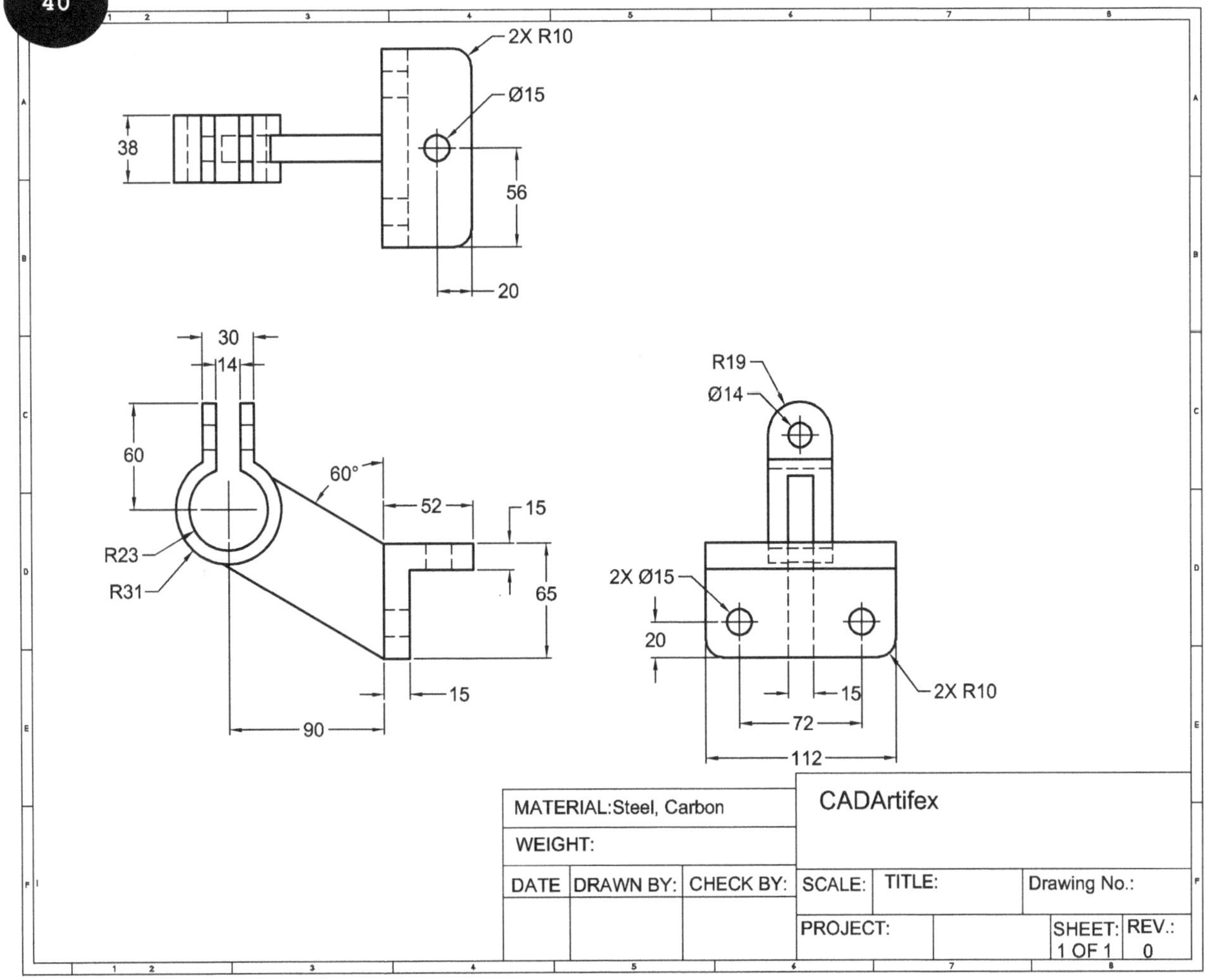

2X R10

Ø15

38

56

20

30

14

60

60°

52

15

R23

65

R31

15

90

R19

Ø14

2X Ø15

20

15

2X R10

72

112

MATERIAL:Steel, Carbon			CADArtifex			
WEIGHT:						
DATE	DRAWN BY:	CHECK BY:	SCALE:	TITLE:		Drawing No.:
			PROJECT:			SHEET: 1 OF 1	REV.: 0

Exercise 21.

Create the 3D model, as shown in Figure 41. Different views of the model and dimensions are shown in Figure 42. After creating the model, assign the Stainless Steel AISI 304 material and calculate its mass properties. All dimensions are in mm.

41

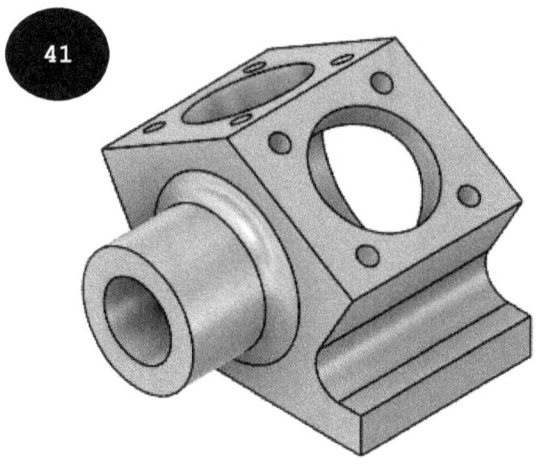

42

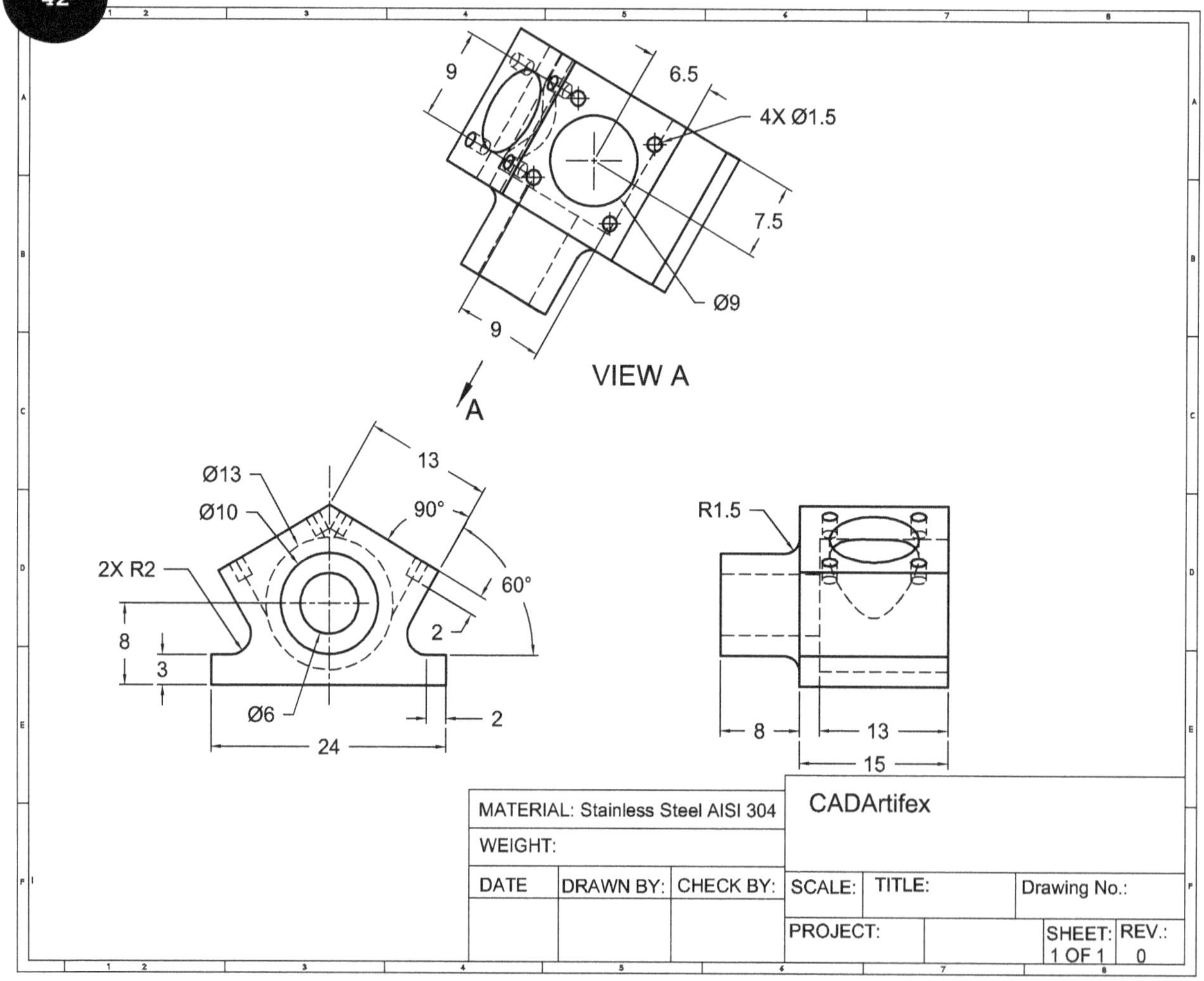

VIEW A

9   6.5   4X Ø1.5

7.5

Ø9

9

A

Ø13   13   90°

Ø10

2X R2   60°

8   2

3

Ø6   2

24

R1.5

8   13

15

MATERIAL: Stainless Steel AISI 304			CADArtifex		
WEIGHT:					
DATE	DRAWN BY:	CHECK BY:	SCALE:	TITLE:	Drawing No.:
			PROJECT:		SHEET: 1 OF 1 / REV.: 0

Exercise 22.

Create the 3D model, as shown in Figure 43. Different views of the model and dimensions are shown in Figure 44. After creating the model, assign the Steel AISI 1020 107 HR material and calculate its mass properties. All dimensions are in mm.

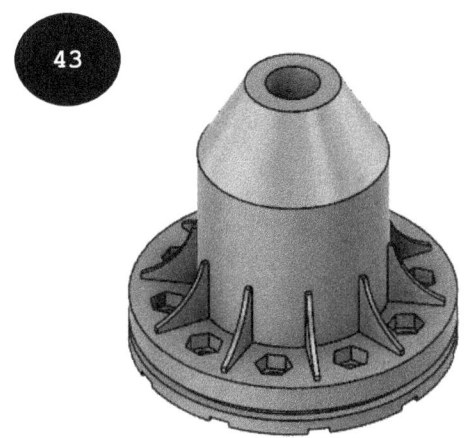

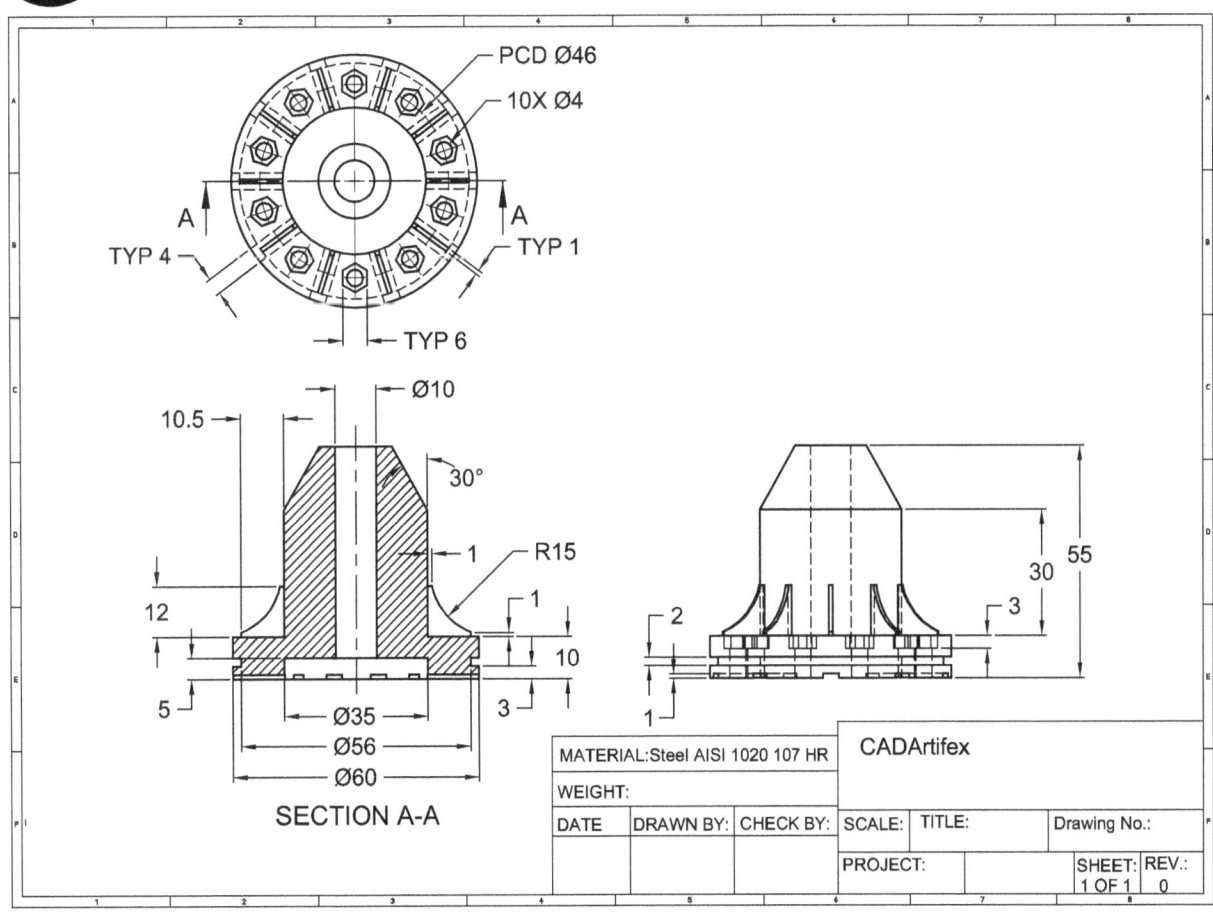

Exercise 23.

Create the 3D model, as shown in Figure 45. Different views of the model and dimensions are shown in Figure 46. After creating the model, assign the Stainless Steel material and calculate its mass properties. All dimensions are in mm.

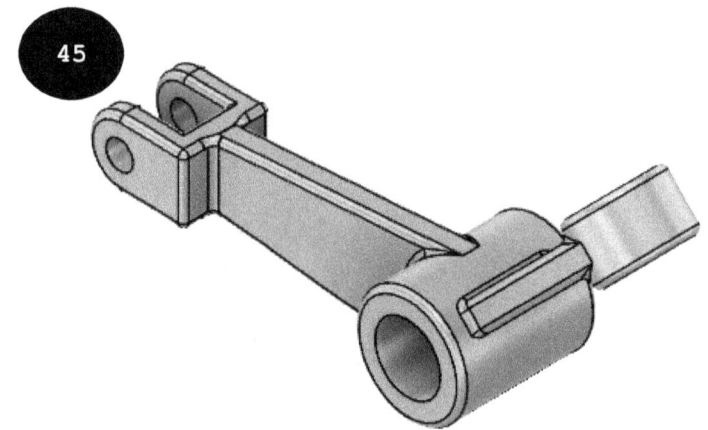

45

46

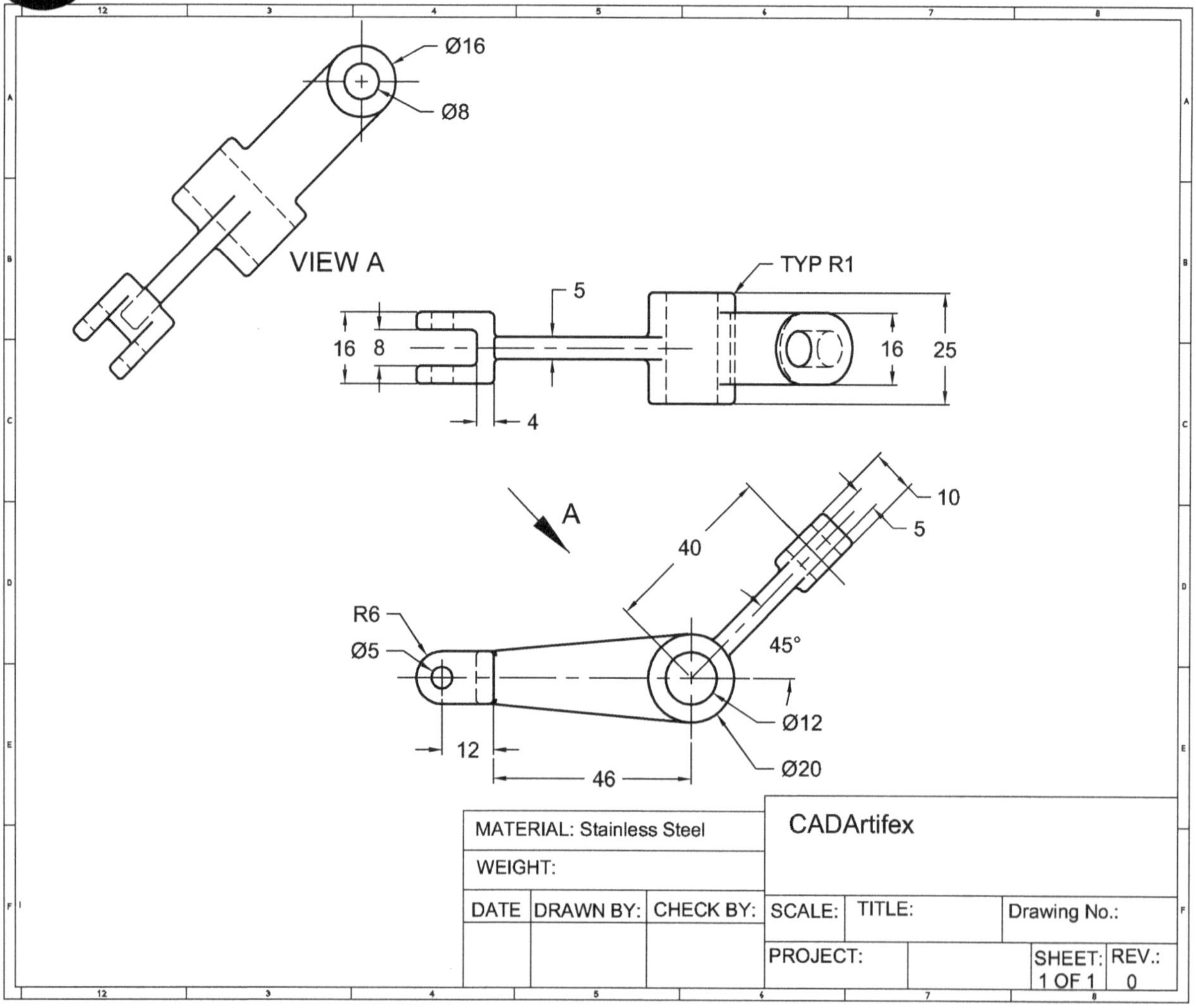

VIEW A

Ø16
Ø8

TYP R1

5

16 8

4

16 25

A

40

10
5

45°

R6
Ø5

12

46

Ø12

Ø20

MATERIAL: Stainless Steel			CADArtifex			
WEIGHT:						
DATE	DRAWN BY:	CHECK BY:	SCALE:	TITLE:		Drawing No.:
			PROJECT:			SHEET: REV.:
						1 OF 1 0

Exercise 24.

Create the 3D model, as shown in Figure 47. Different views of the model and dimensions are shown in Figure 48. After creating the model, assign the Steel AISI 1020 107 HR material and calculate its mass properties. All dimensions are in mm.

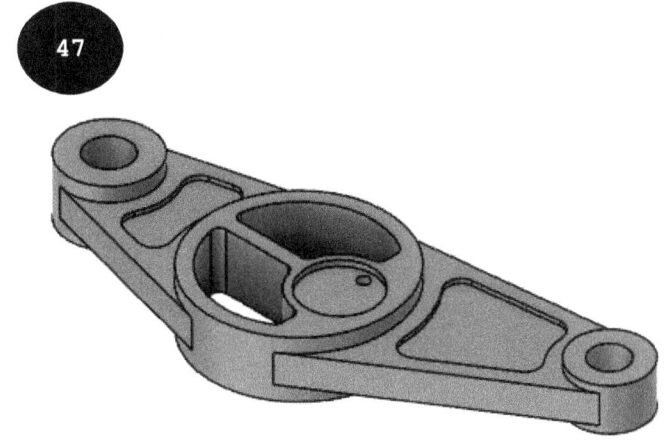

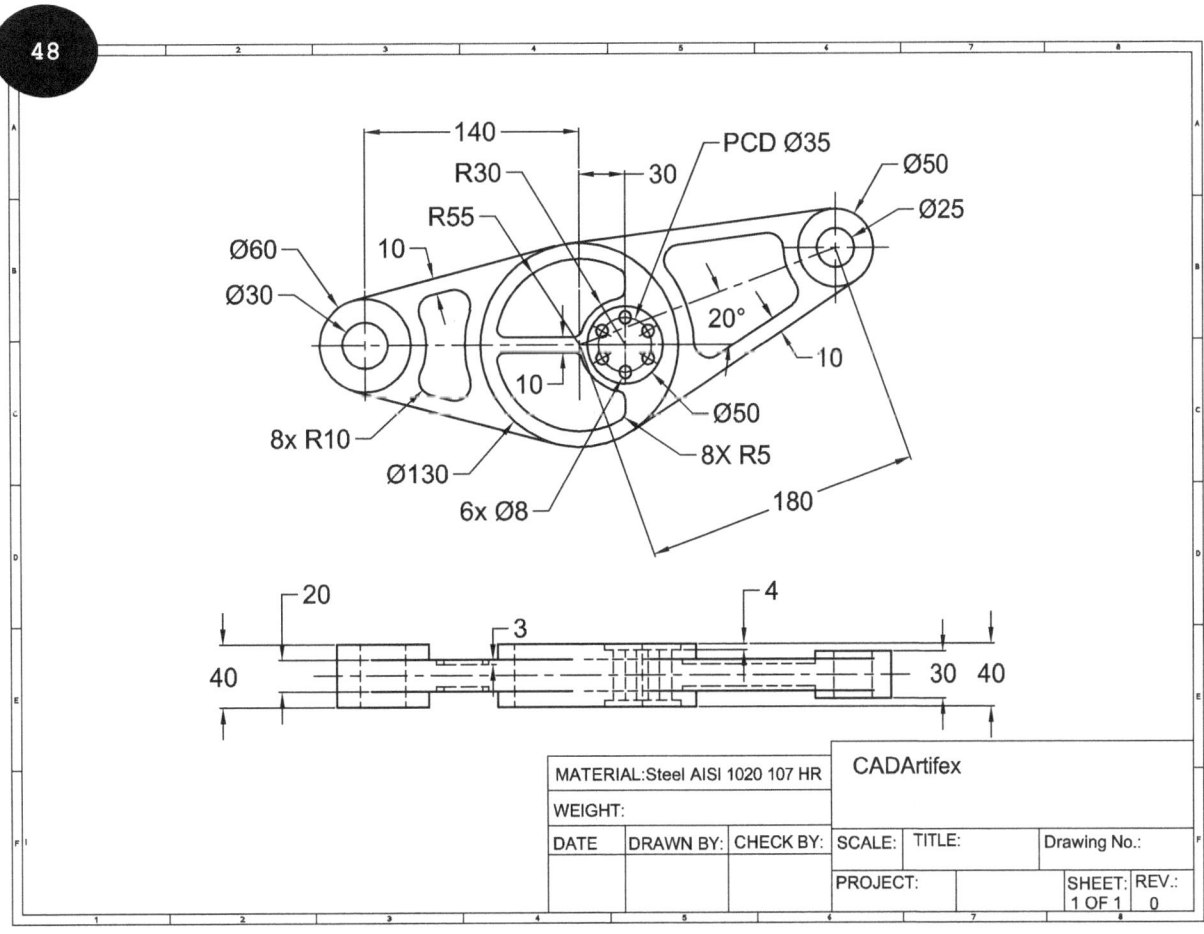

Exercise 25.

Create the 3D model, as shown in Figure 49. Different views of the model and dimensions are shown in Figure 50. After creating the model, assign the Stainless Steel material and calculate its mass properties. All dimensions are in mm.

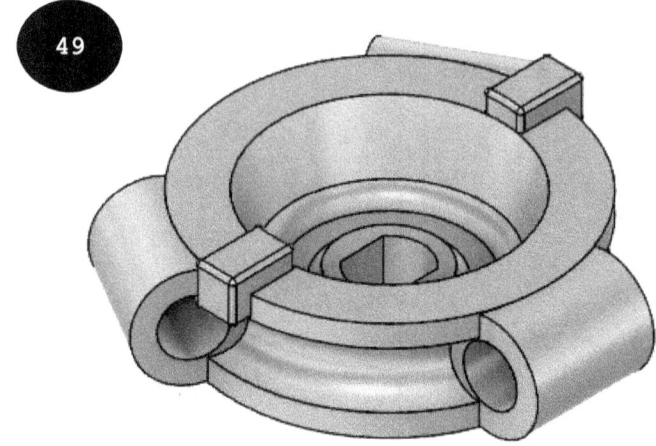

49

50

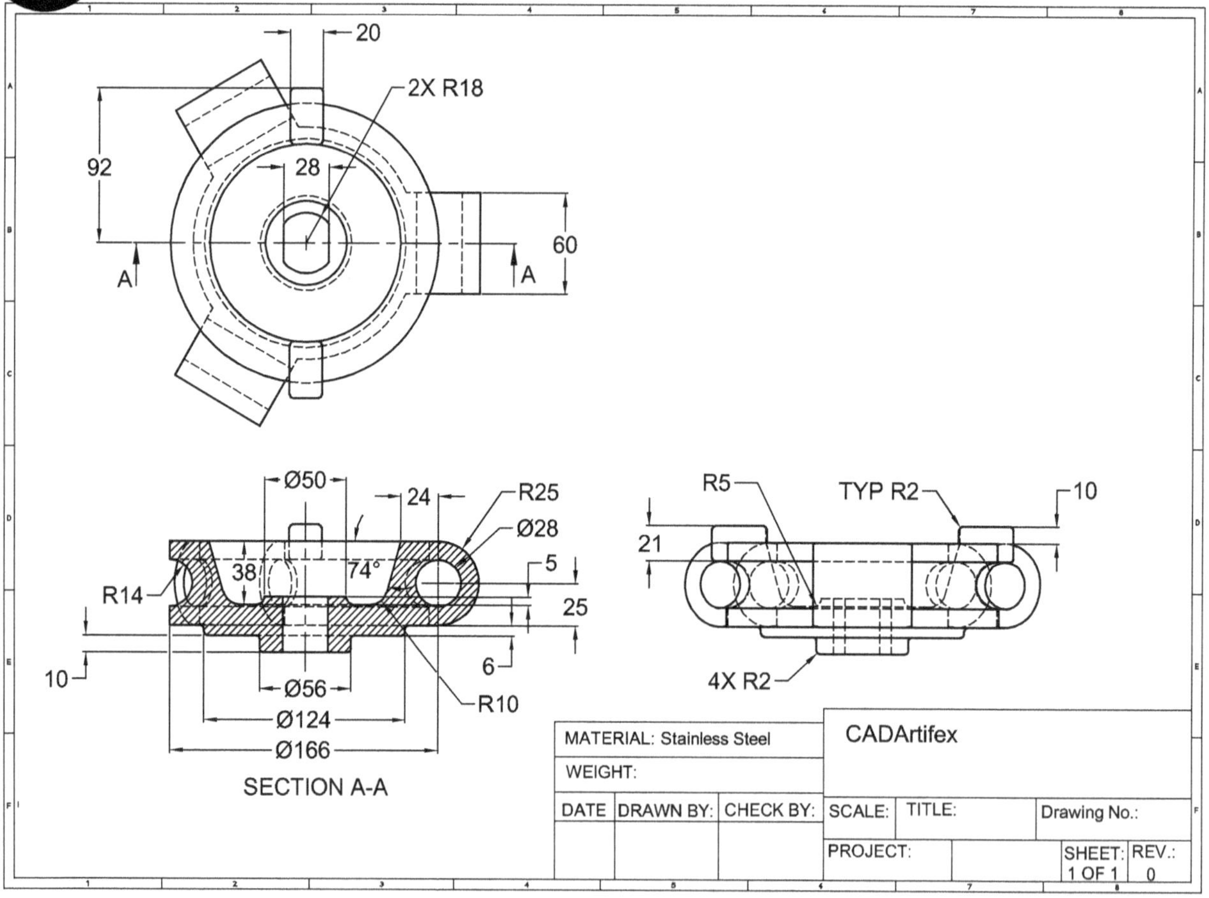

20
2X R18
92
28
60
A
A
Ø50
24
R25
Ø28
38
74°
5
R14
25
10
6
Ø56
R10
Ø124
Ø166
SECTION A-A
R5
TYP R2
10
21
4X R2
MATERIAL: Stainless Steel
WEIGHT:
CADArtifex
DATE
DRAWN BY:
CHECK BY:
SCALE:
TITLE:
Drawing No.:
PROJECT:
SHEET:
1 OF 1
REV.:
0

Exercise 26.

Create the 3D model, as shown in Figure 51. Different views of the model and dimensions are shown in Figure 52. After creating the model, assign the Stainless Steel AISI 304 material and calculate its mass properties. All dimensions are in mm.

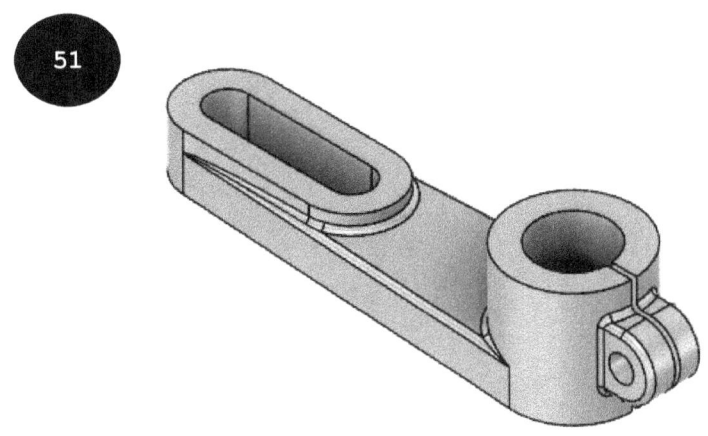

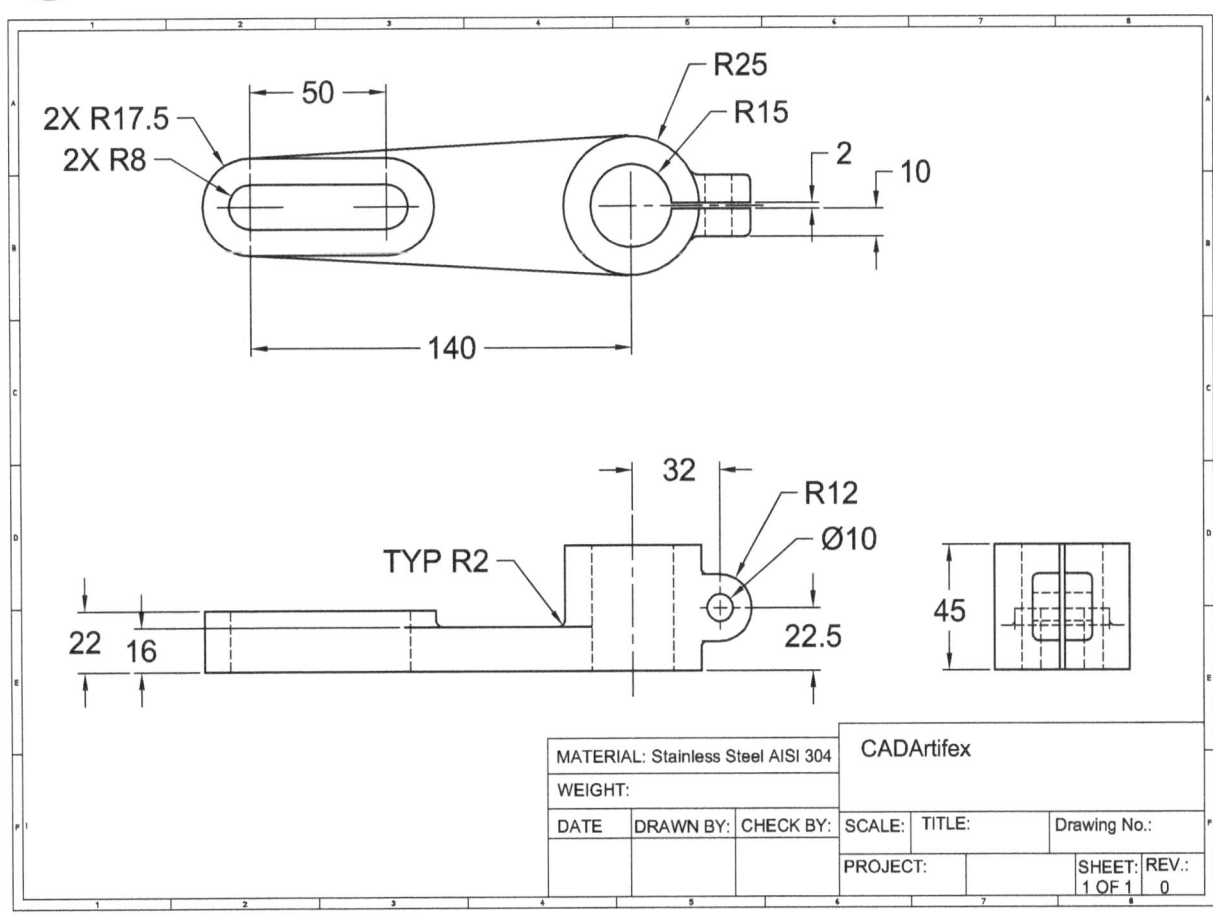

Exercise 27.

Create the 3D model, as shown in Figure 53. Different views of the model and dimensions are shown in Figure 54. After creating the model, assign the Steel AISI 1020 107 HR material and calculate its mass properties. All dimensions are in mm.

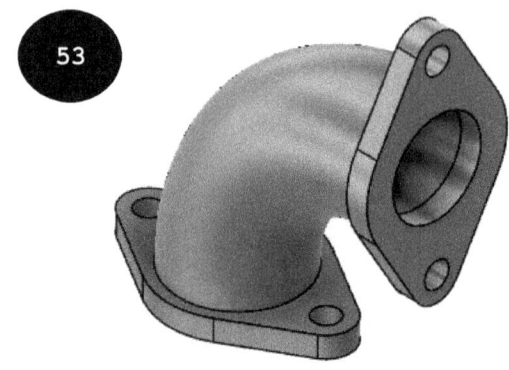

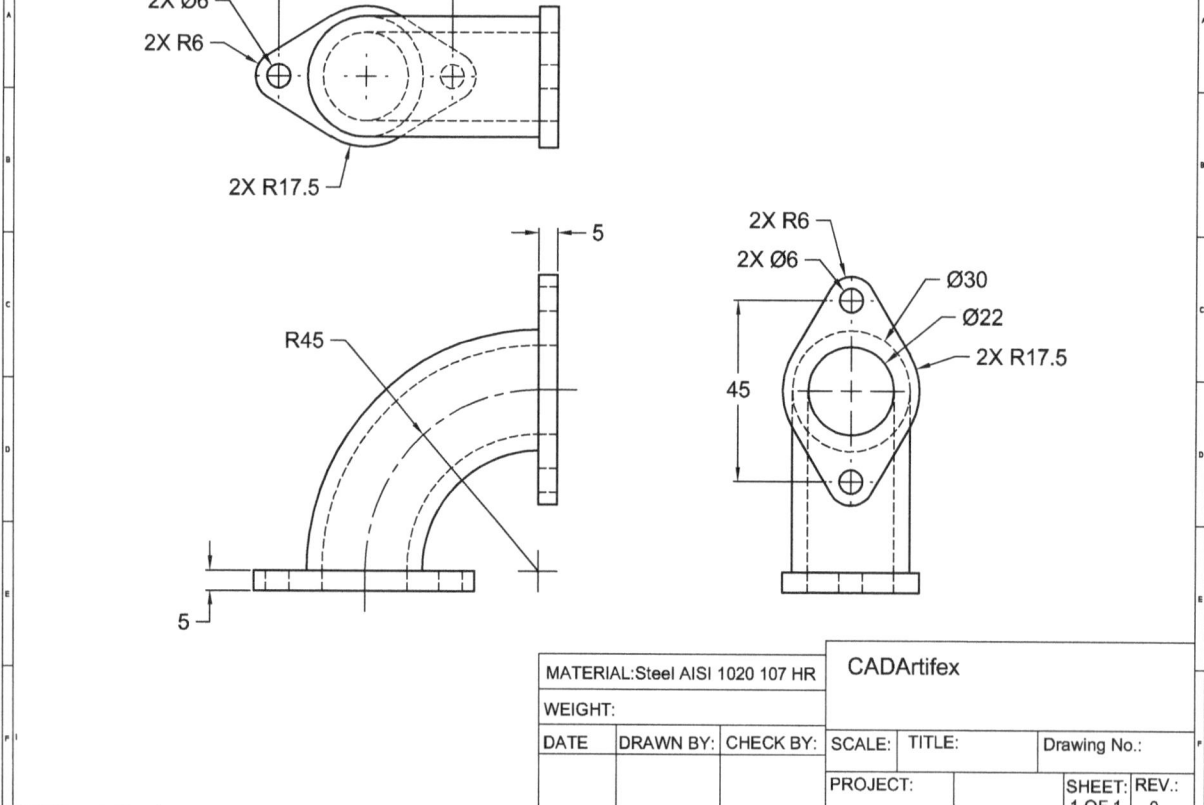

Exercise 28.

Create the 3D model, as shown in Figure 55. Different views of the model and dimensions are shown in Figure 56. After creating the model, assign the Stainless Steel material and calculate its mass properties. All dimensions are in mm.

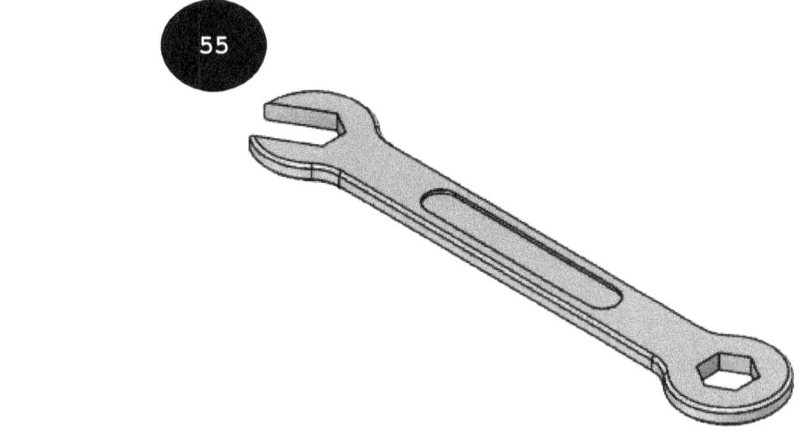

55

56

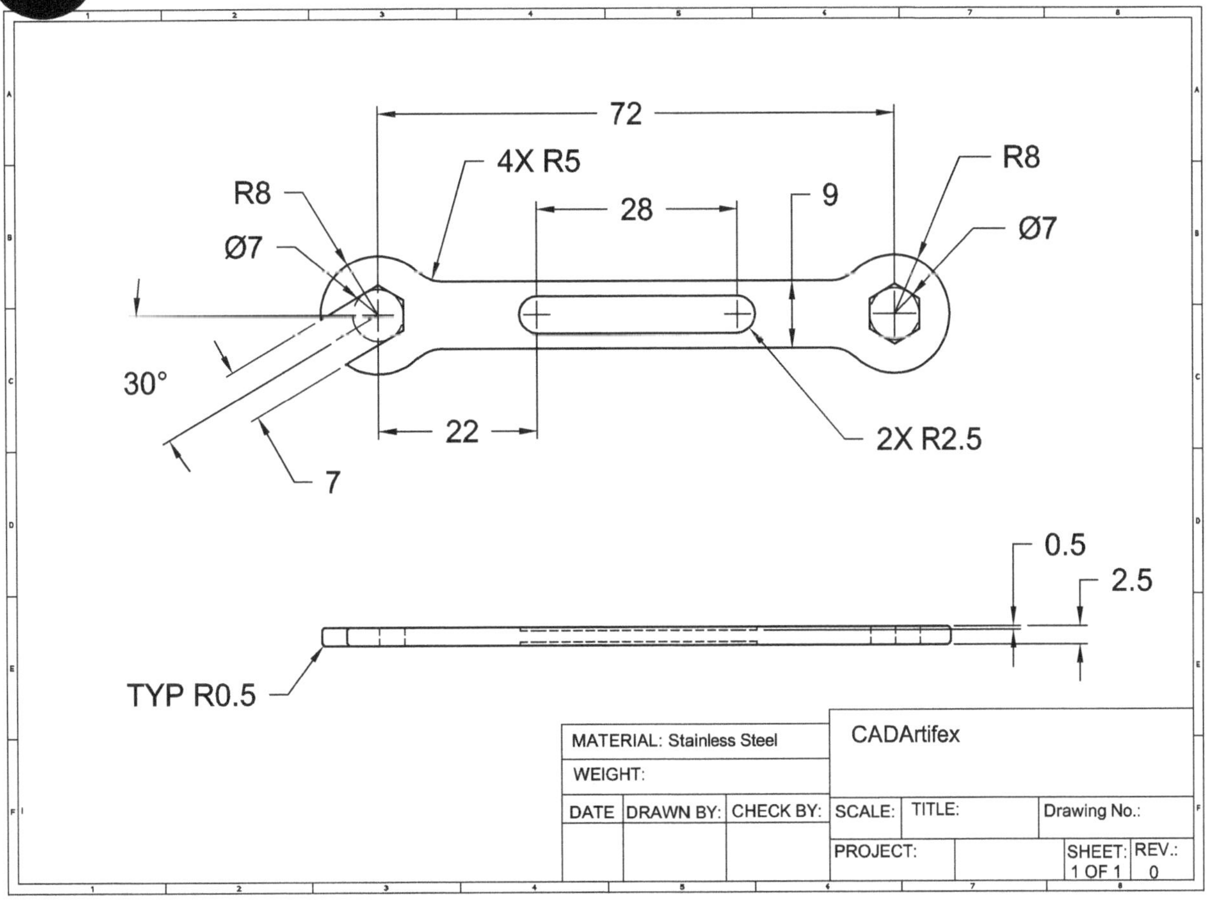

Exercise 29.

Create the 3D model, as shown in Figure 57. Different views of the model and dimensions are shown in Figure 58. After creating the model, assign the Platinum and calculate its mass properties. All dimensions are in mm.

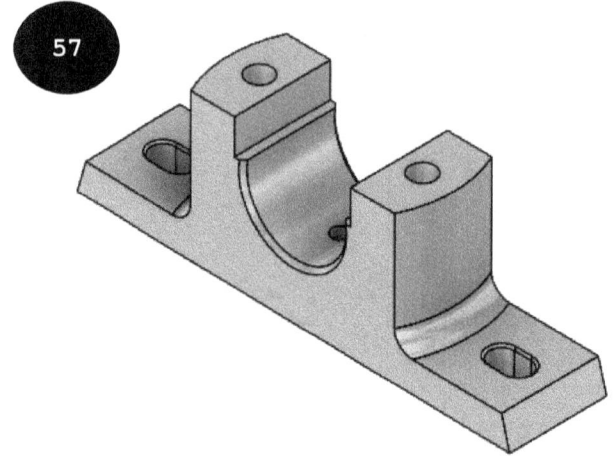

57

58

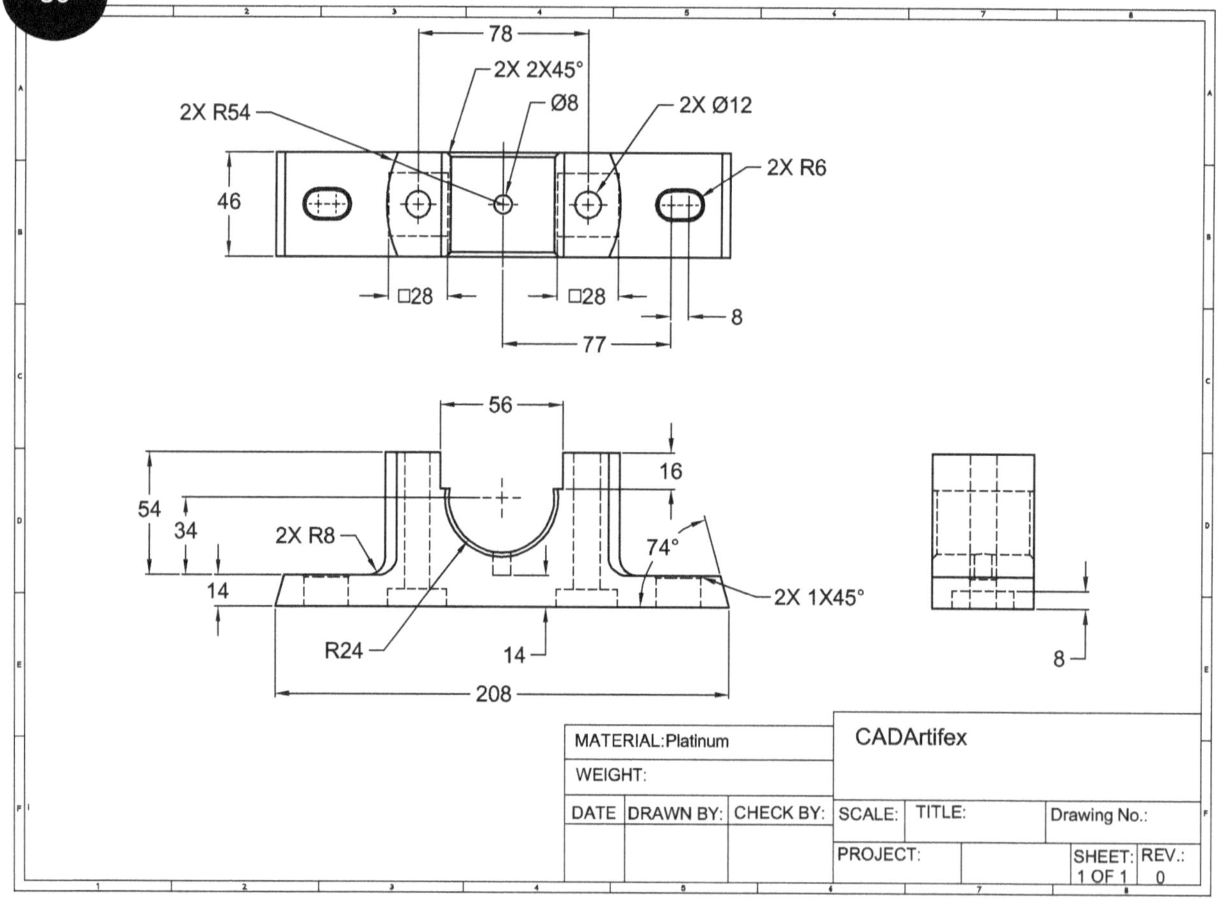

2X R54 — 78 — 2X 2X45° — Ø8 — 2X Ø12 — 2X R6

46

□28 □28 8

— 77 —

56

54 34 2X R8 16

14 74° 2X 1X45°

R24 14

— 208 —

8

MATERIAL:Platinum			CADArtifex		
WEIGHT:					
DATE	DRAWN BY:	CHECK BY:	SCALE:	TITLE:	Drawing No.:
			PROJECT:		SHEET: 1 OF 1 / REV.: 0

Exercise 30.

Create the 3D model, as shown in Figure 59. Different views of the model and dimensions are shown in Figure 60. After creating the model, assign the Stainless Steel AISI 304 material and calculate its mass properties. All dimensions are in mm.

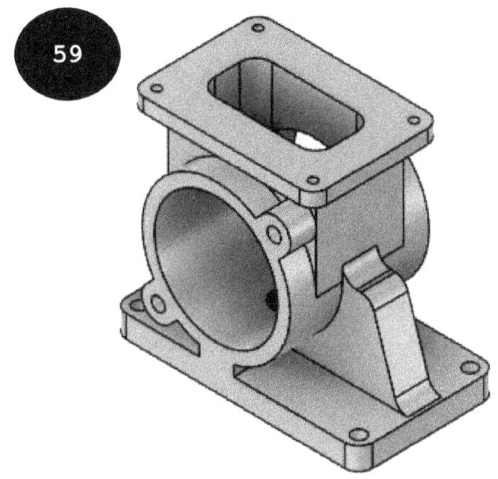

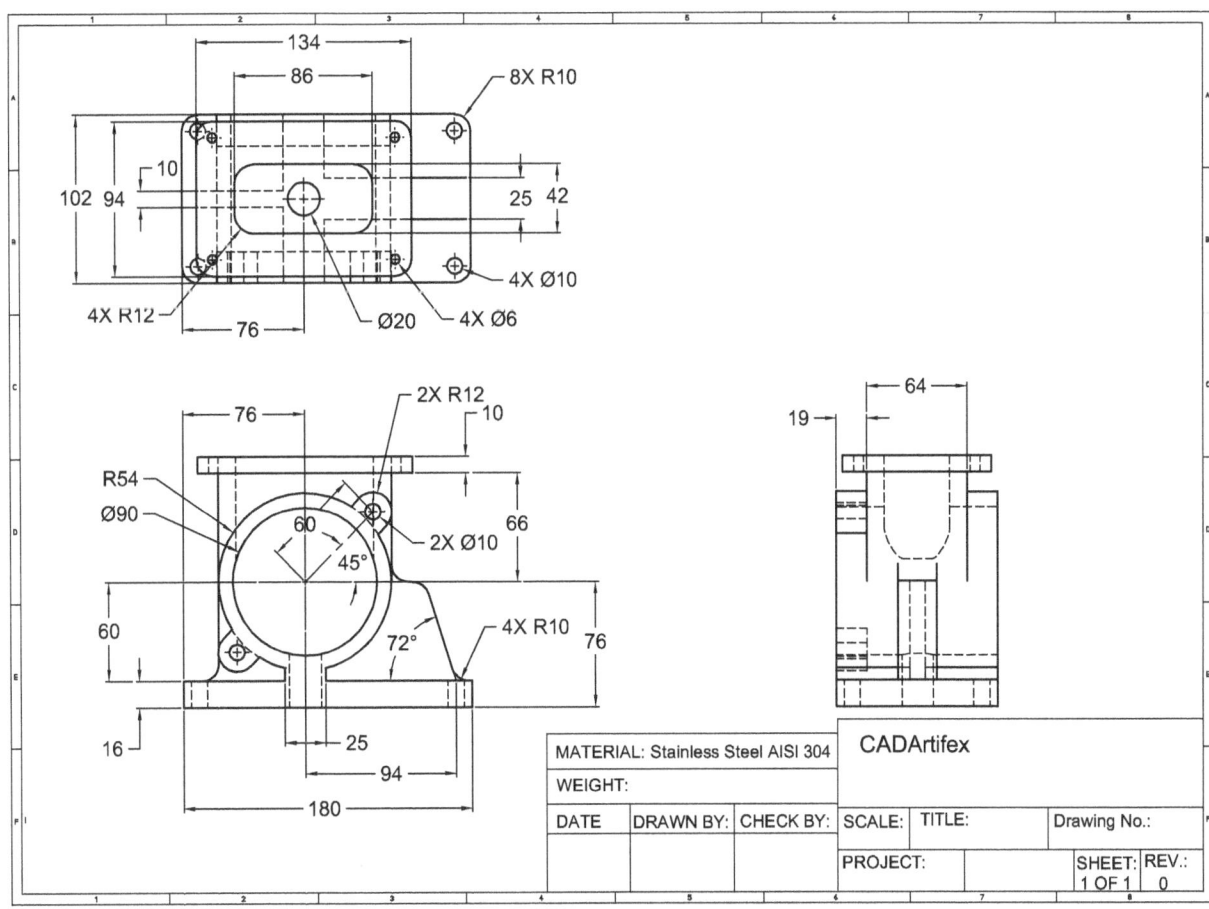

Exercise 31.

Create the 3D model, as shown in Figure 61. Different views of the model and dimensions are shown in Figure 62. After creating the model, assign the Steel AISI 1020 107 HR material and calculate its mass properties. All dimensions are in mm.

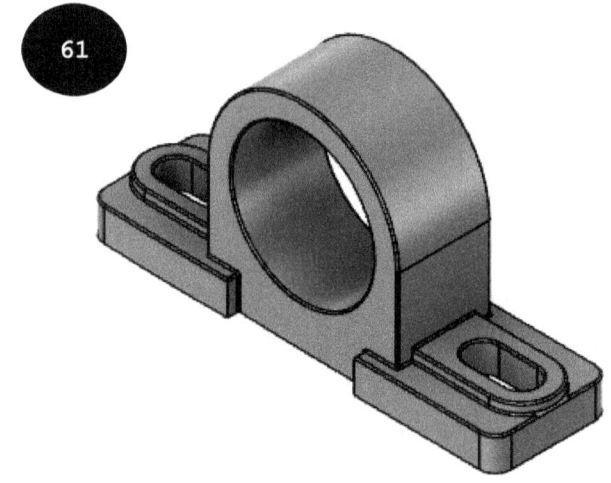

61

62

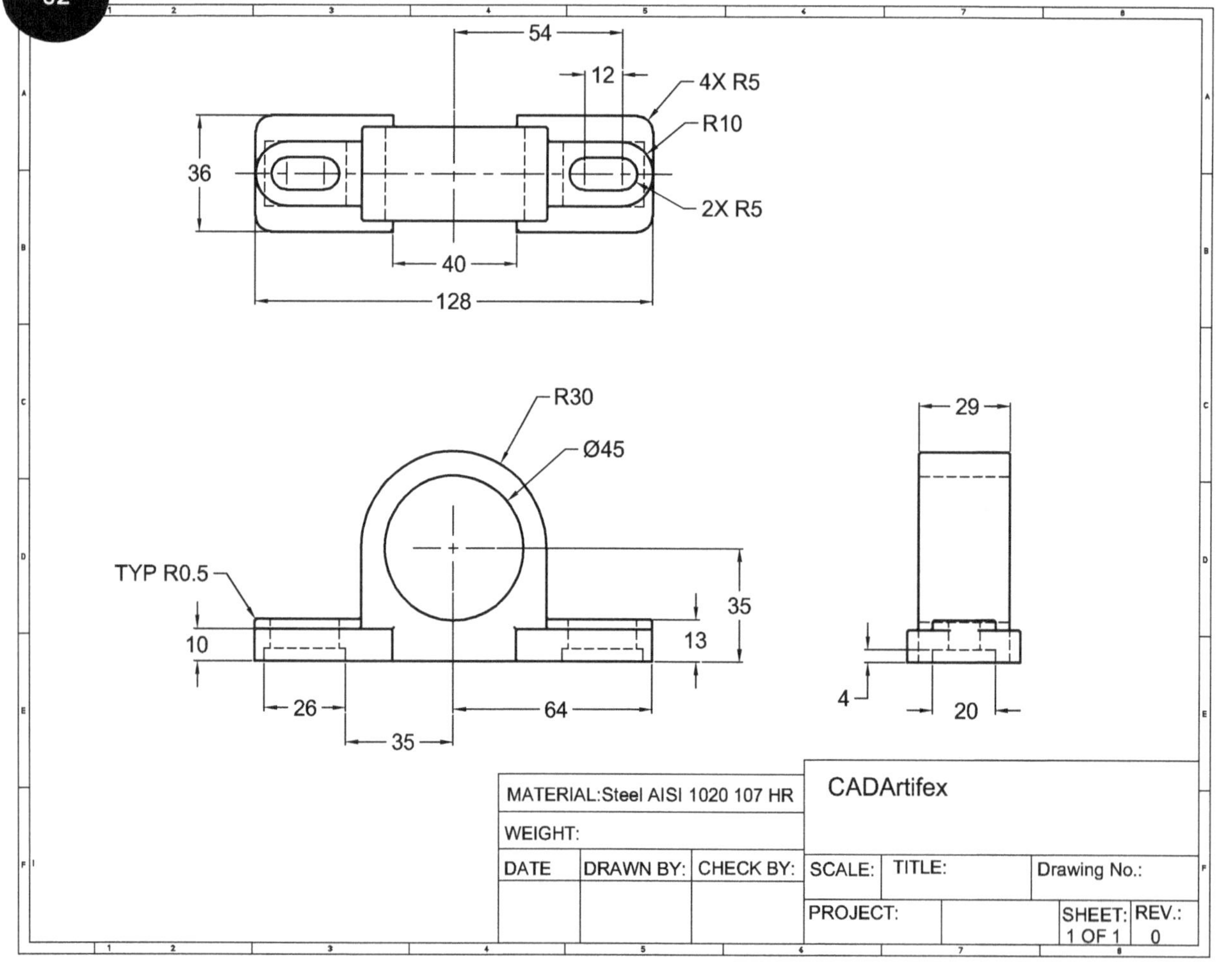

54

12

4X R5

R10

36

2X R5

40

128

R30

Ø45

TYP R0.5

29

10

35

13

26

64

35

4

20

MATERIAL:Steel AISI 1020 107 HR			CADArtifex				
WEIGHT:							
DATE	DRAWN BY:	CHECK BY:	SCALE:	TITLE:		Drawing No.:	
			PROJECT:			SHEET: 1 OF 1	REV.: 0

Exercise 32.

Create the 3D model, as shown in Figure 63. Different views of the model and dimensions are shown in Figure 64. After creating the model, assign the Stainless Steel material and calculate its mass properties. All dimensions are in mm.

63

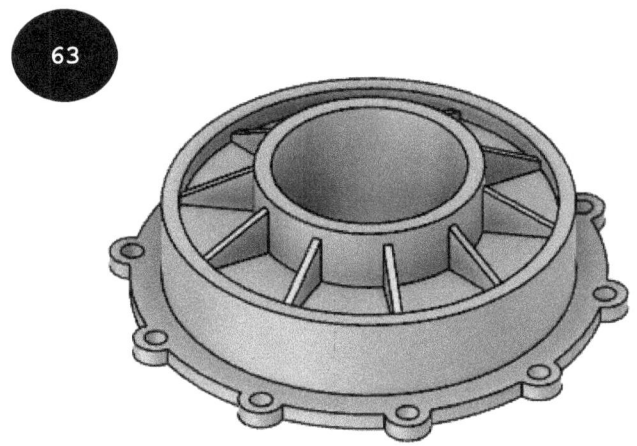

64

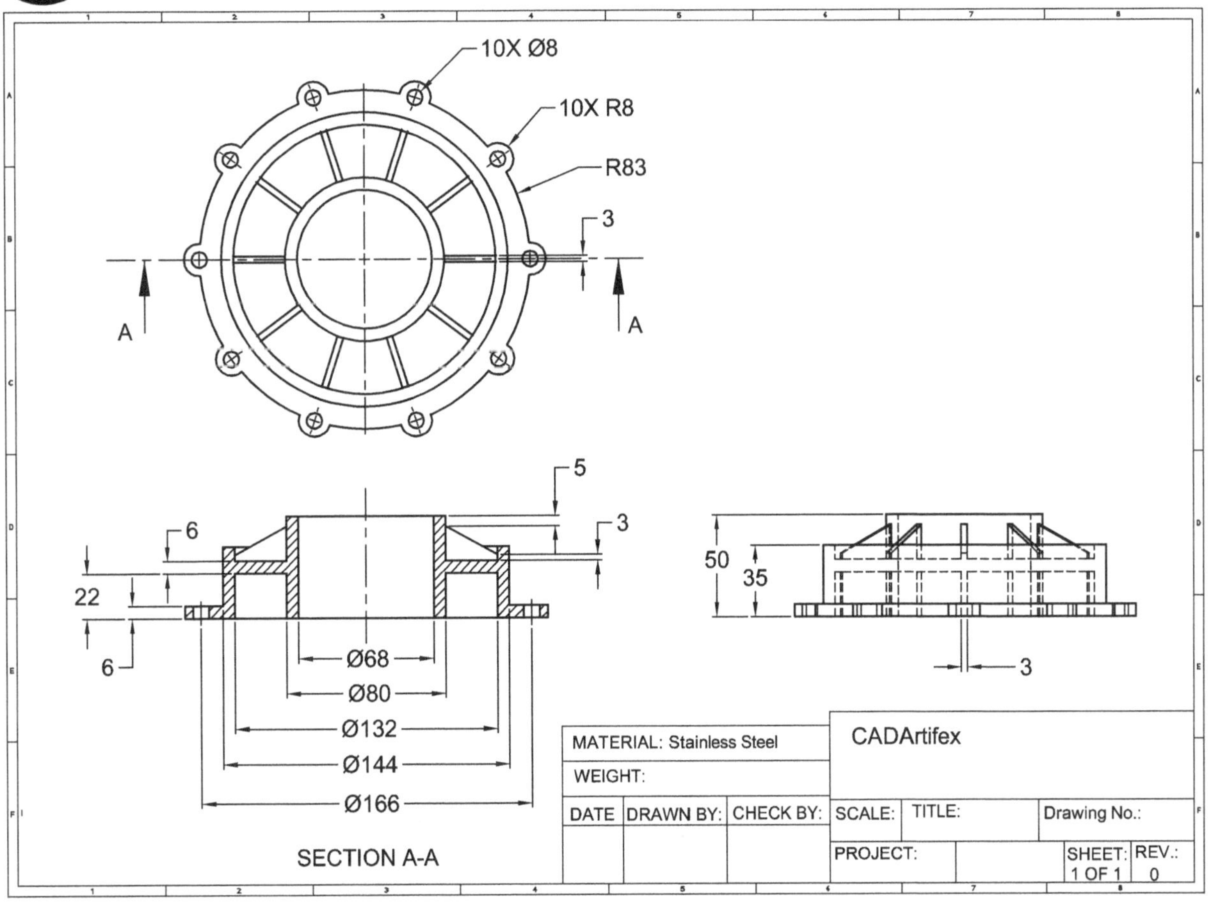

10X Ø8

10X R8

R83

3

A — A

5

3

6

22

6

Ø68

Ø80

Ø132

Ø144

Ø166

50

35

3

SECTION A-A

MATERIAL: Stainless Steel			CADArtifex		
WEIGHT:					
DATE	DRAWN BY:	CHECK BY:	SCALE:	TITLE:	Drawing No.:
			PROJECT:		SHEET: 1 OF 1 / REV.: 0

Exercise 33.

Create the 3D model, as shown in Figure 65. Different views of the model and dimensions are shown in Figure 66. After creating the model, assign the Platinum material and calculate its mass properties. All dimensions are in mm.

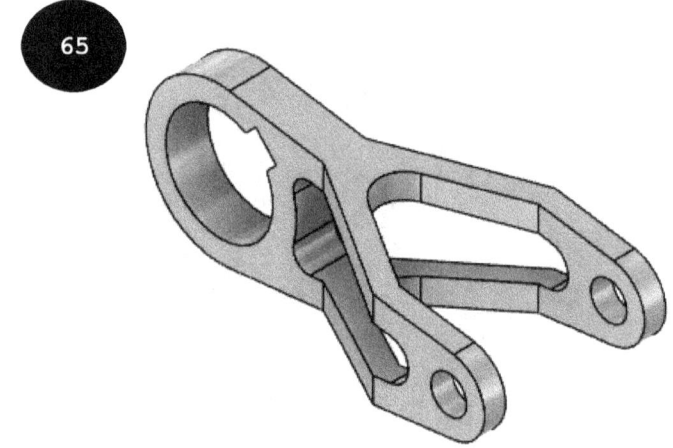

65

66

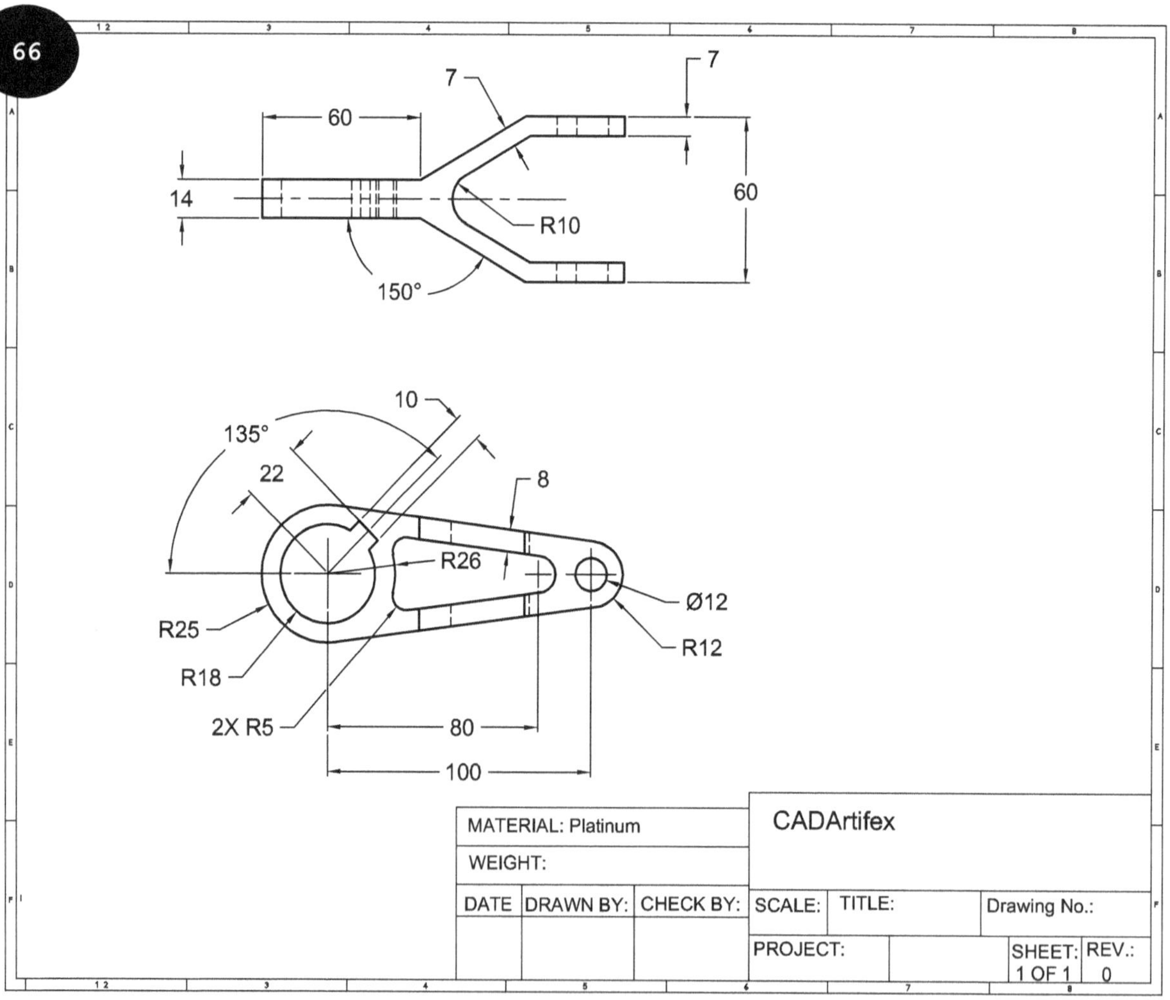

7

60

7

14

60

R10

150°

135°

10

22

8

R26

R25

Ø12

R18

R12

2X R5

80

100

MATERIAL: Platinum			CADArtifex		
WEIGHT:					
DATE	DRAWN BY:	CHECK BY:	SCALE:	TITLE:	Drawing No.:
			PROJECT:		SHEET: 1 OF 1 / REV.: 0

Exercise 34.

Create the 3D model, as shown in Figure 67. Different views of the model and dimensions are shown in Figure 68. After creating the model, assign the Stainless Steel AISI 304 material and calculate its mass properties. All dimensions are in mm.

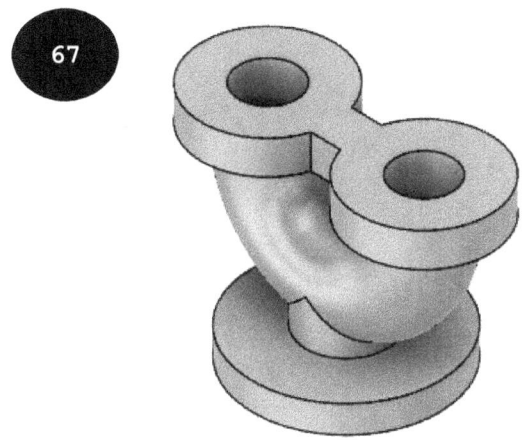

67

68

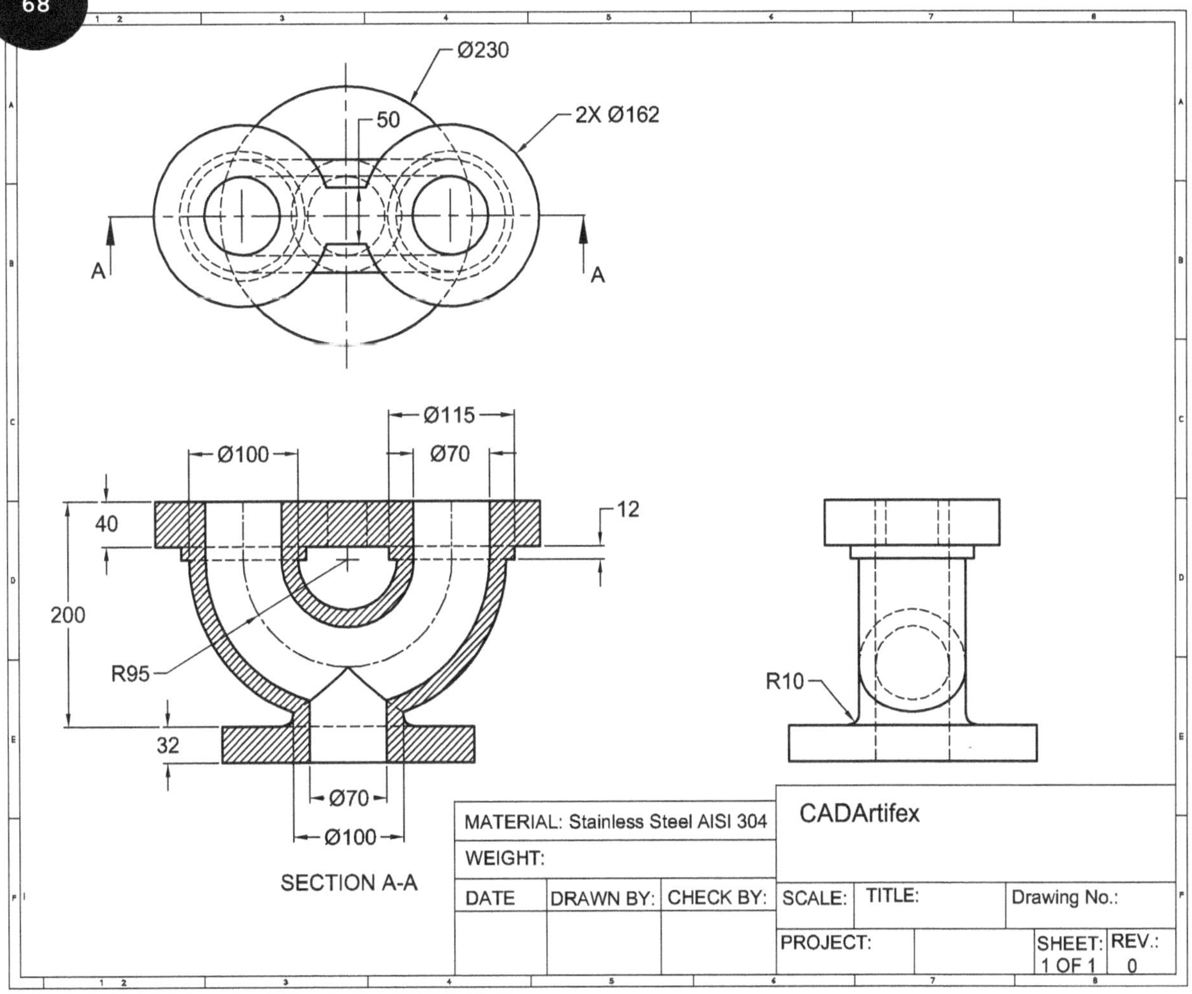

Ø230

50

2X Ø162

A — A

Ø115

Ø100

Ø70

12

40

200

R95

32

R10

Ø70

Ø100

SECTION A-A

MATERIAL: Stainless Steel AISI 304			CADArtifex		
WEIGHT:					
DATE	DRAWN BY:	CHECK BY:	SCALE:	TITLE:	Drawing No.:
			PROJECT:		SHEET: 1 OF 1 · REV.: 0

Exercise 35.

Create the 3D model, as shown in Figure 69. Different views of the model and dimensions are shown in Figure 70. After creating the model, assign the Steel AISI 1020 107 HR material and calculate its mass properties. All dimensions are in mm.

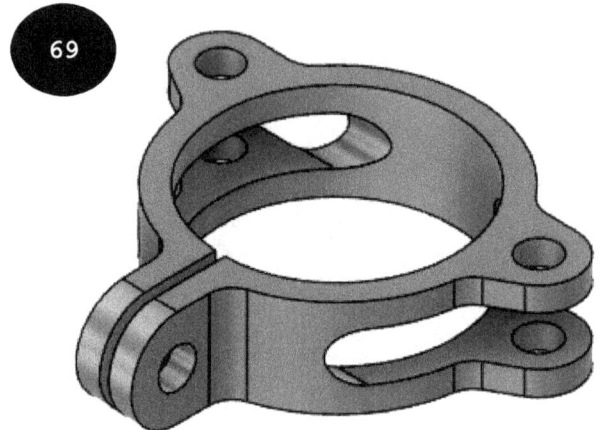

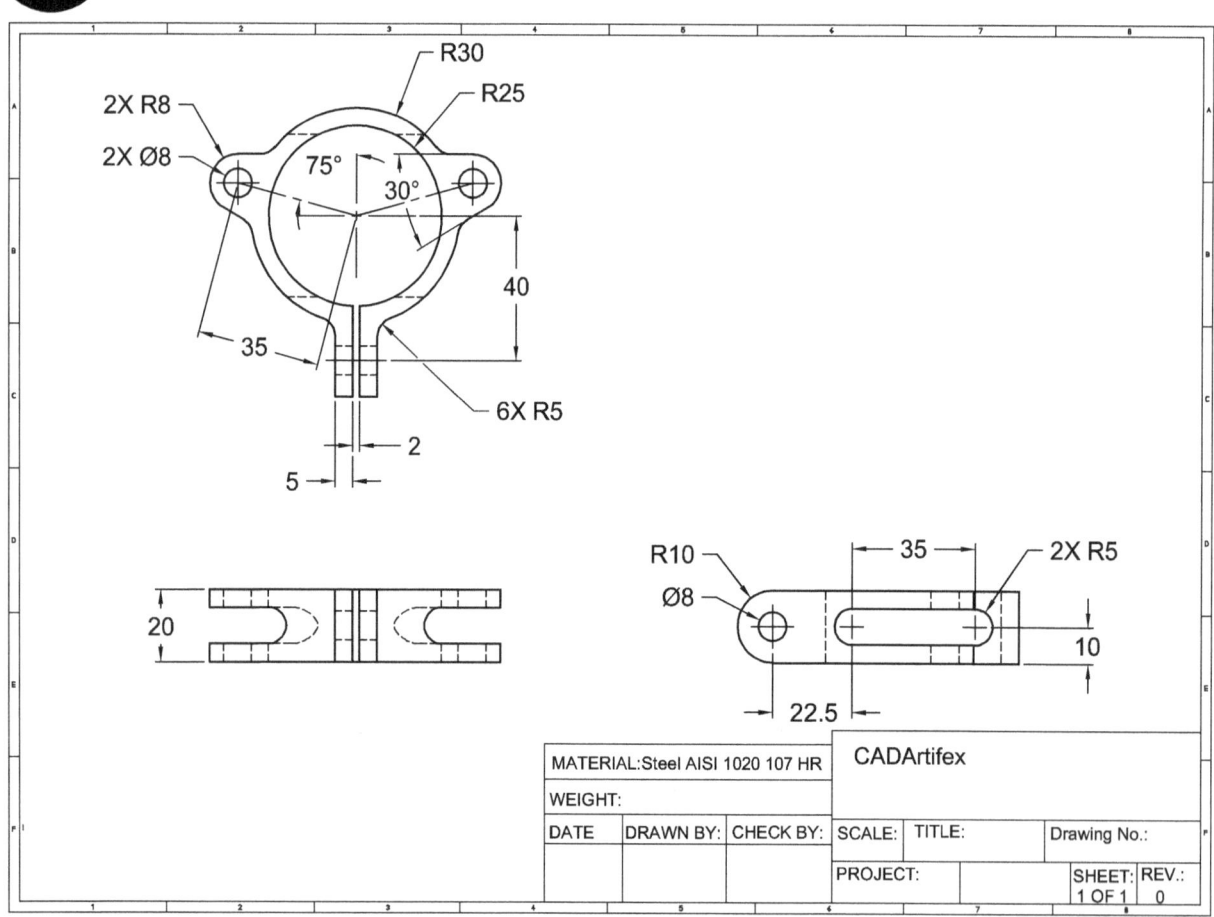

Exercise 36.

Create the 3D model, as shown in Figure 71. Different views of the model and dimensions are shown in Figure 72. After creating the model, assign the Stainless Steel material and calculate its mass properties. All dimensions are in mm.

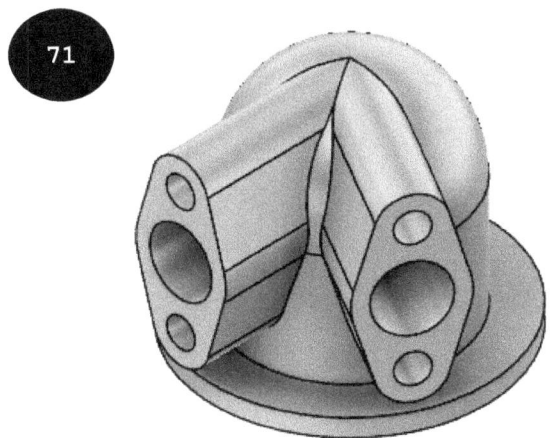

71

72

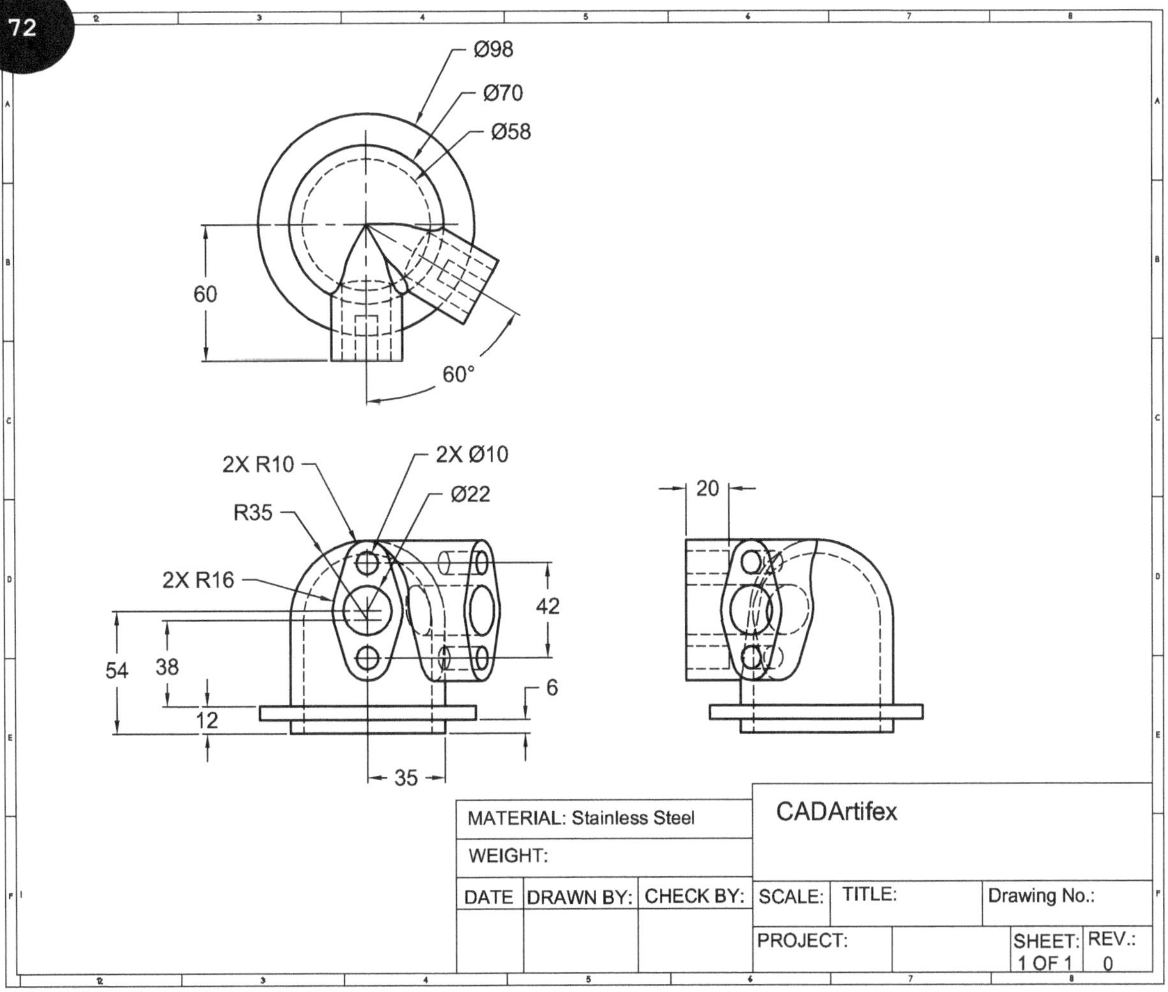

Ø98
Ø70
Ø58

60

60°

2X R10
2X Ø10
R35
Ø22
2X R16
42
54
38
12
6
35
20

MATERIAL: Stainless Steel

WEIGHT:

CADArtifex

DATE | DRAWN BY: | CHECK BY: | SCALE: | TITLE: | Drawing No.:

PROJECT: | | SHEET: 1 OF 1 | REV.: 0

Exercise 37.

Create the 3D model, as shown in Figure 73. Different views of the model and dimensions are shown in Figure 74. After creating the model, assign the Platinum material and calculate its mass properties. All dimensions are in mm.

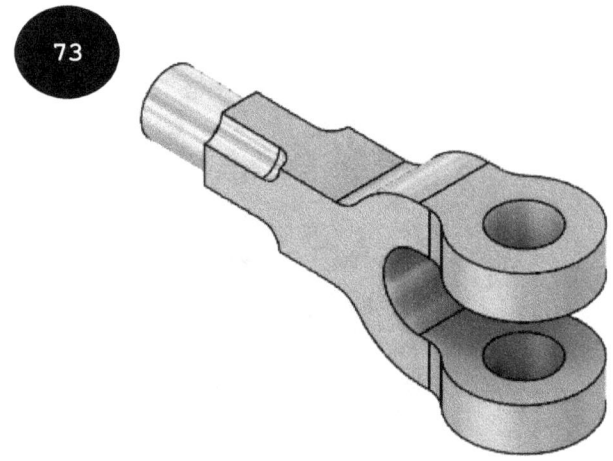

73

74

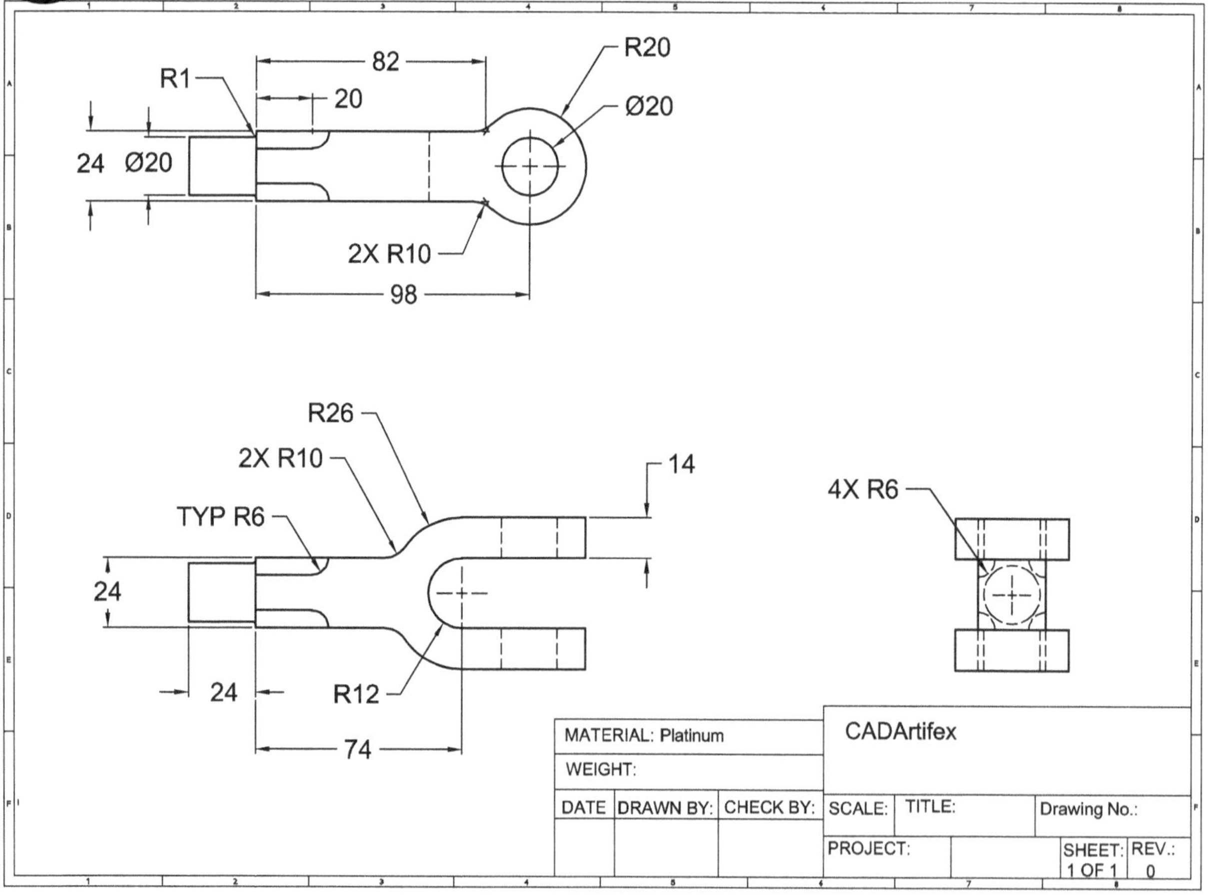

R20
82
R1
20
Ø20
24 Ø20
2X R10
98

R26
2X R10
14
TYP R6
4X R6
24
24
R12
74

MATERIAL: Platinum			CADArtifex		
WEIGHT:					
DATE	DRAWN BY:	CHECK BY:	SCALE:	TITLE:	Drawing No.:
			PROJECT:		SHEET: 1 OF 1	REV.: 0

Exercise 38.

Create the 3D model, as shown in Figure 75. Different views of the model and dimensions are shown in Figure 76. After creating the model, assign the Stainless Steel material and calculate its mass properties. All dimensions are in mm.

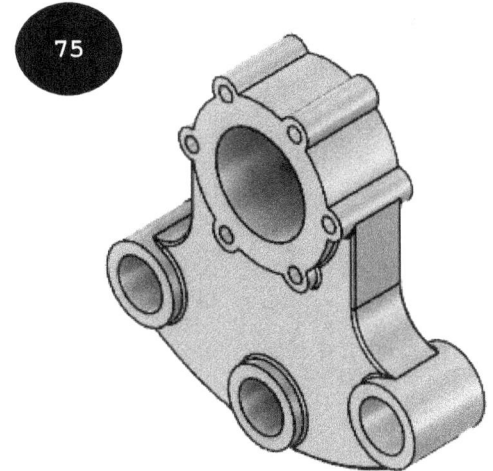

75

76

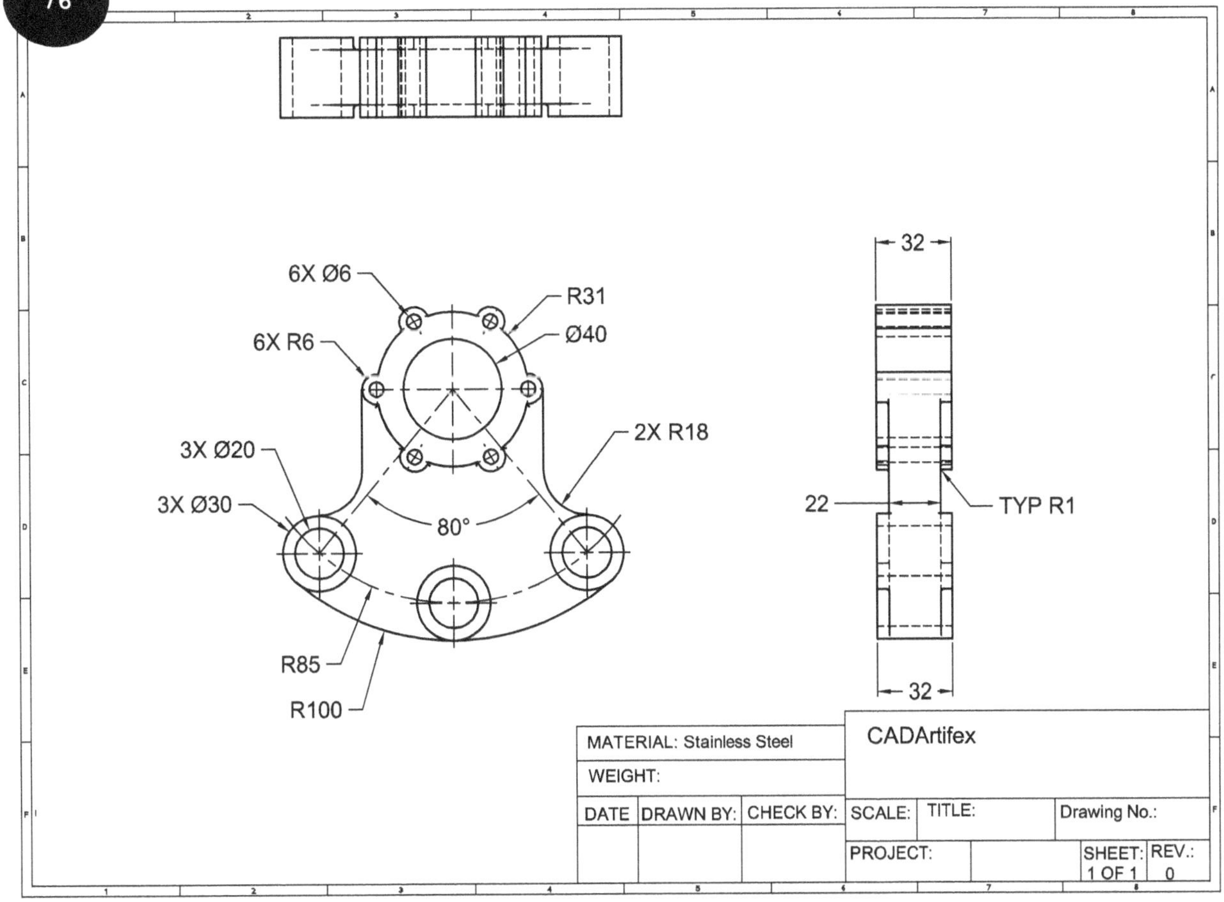

6X Ø6

R31

6X R6

Ø40

3X Ø20

2X R18

3X Ø30

80°

22

TYP R1

R85

R100

32

32

MATERIAL: Stainless Steel	CADArtifex			
WEIGHT:				
DATE	DRAWN BY:	CHECK BY:	SCALE:	TITLE:	Drawing No.:
			PROJECT:		SHEET: REV.: 1 OF 1 0

Exercise 39.

Create the 3D model, as shown in Figure 77. Different views of the model and dimensions are shown in Figure 78. After creating the model, assign the Platinum material and calculate its mass properties. All dimensions are in mm.

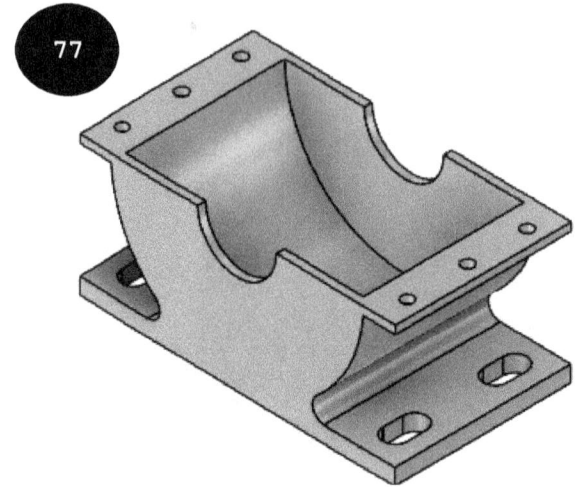

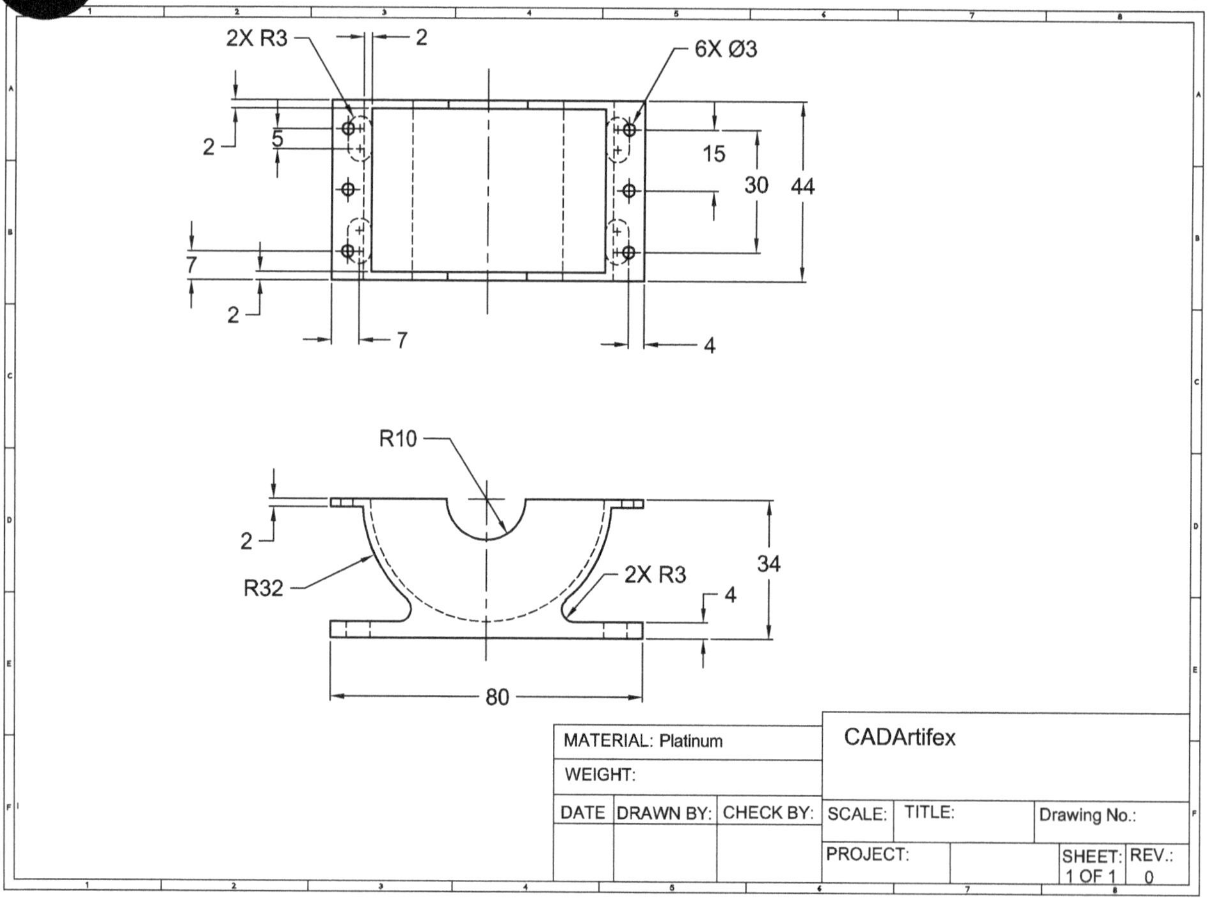

Exercise 40.

Create the 3D model, as shown in Figure 79. Different views of the model and dimensions are shown in Figure 80. After creating the model, assign the Stainless Steel AISI 304 material and calculate its mass properties. All dimensions are in mm.

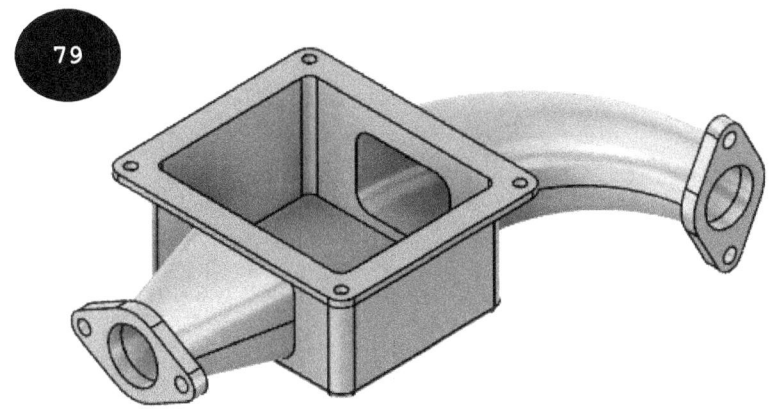

79

80

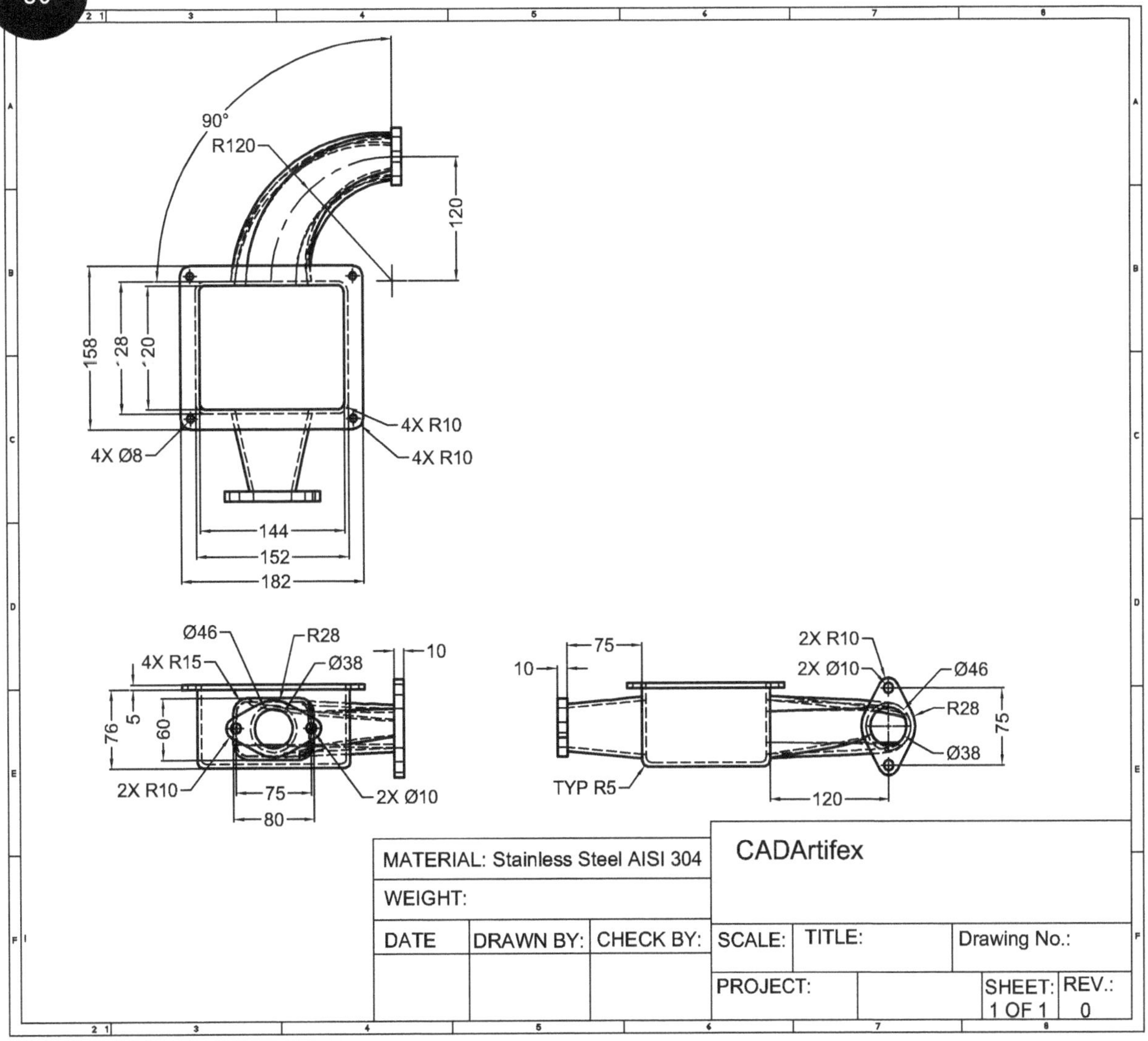

MATERIAL: Stainless Steel AISI 304

WEIGHT:

CADArtifex

DATE	DRAWN BY:	CHECK BY:	SCALE:	TITLE:		Drawing No.:
			PROJECT:			SHEET: 1 OF 1	REV.: 0

Exercise 41.

Create the 3D model, as shown in Figure 81. Different views of the model and dimensions are shown in Figure 82. After creating the model, assign the Steel AISI 1020 107 HR material and calculate its mass properties. All dimensions are in mm.

81

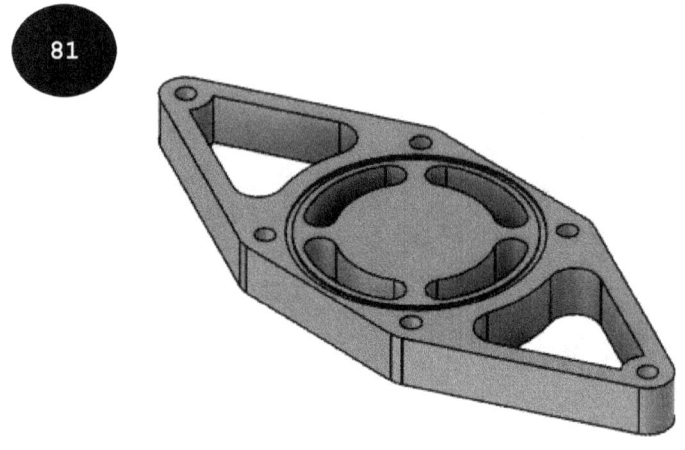

82

Ø95
Ø90
80
2X R6
4X R10
8
6X Ø8
2X R10
R36
60°
R55.5
100
Ø50
3X R10
238
16
1
20

MATERIAL:Steel AISI 1020 107 HR			CADArtifex			
WEIGHT:						
DATE	DRAWN BY:	CHECK BY:	SCALE:	TITLE:	Drawing No.:	
			PROJECT:		SHEET: 1 OF 1	REV.: 0

Exercise 42.

Create the 3D model, as shown in Figure 83. Different views of the model and dimensions are shown in Figure 84. After creating the model, assign the Stainless Steel material and calculate its mass properties. All dimensions are in mm.

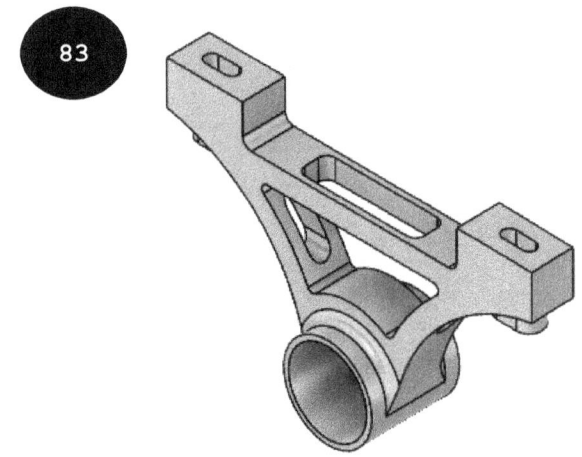

83

84

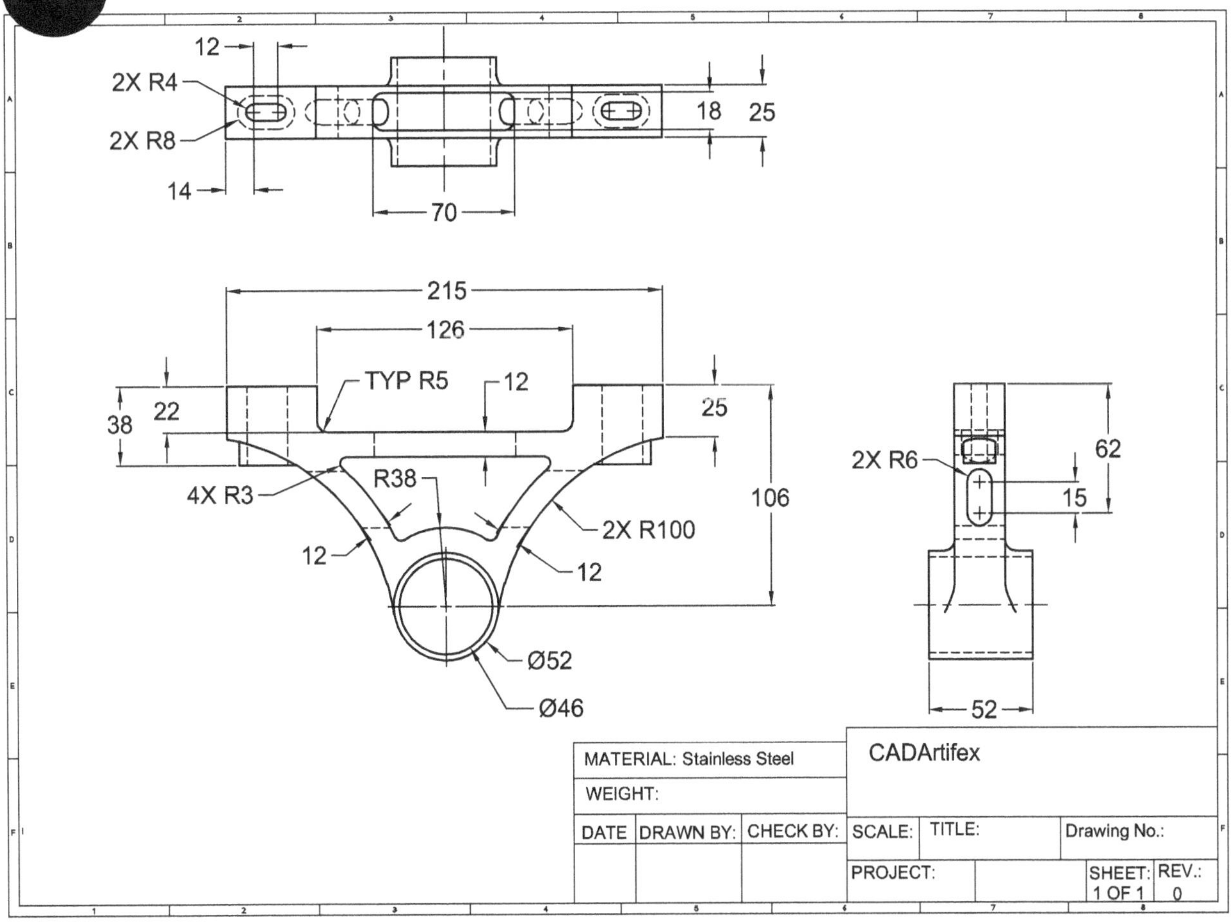

2X R4

2X R8

12

14

18   25

70

2X R4

215

126

TYP R5   12

38   22

25

4X R3   R38

12   2X R100

106

12   12

Ø52

Ø46

2X R6

62

15

52

MATERIAL: Stainless Steel			CADArtifex			
WEIGHT:						
DATE	DRAWN BY:	CHECK BY:	SCALE:	TITLE:		Drawing No.:
			PROJECT:			SHEET: 1 OF 1 / REV.: 0

Exercise 43.

Create the 3D model, as shown in Figure 85. Different views of the model and dimensions are shown in Figure 86. After creating the model, assign the Platinum material and calculate its mass properties. All dimensions are in mm.

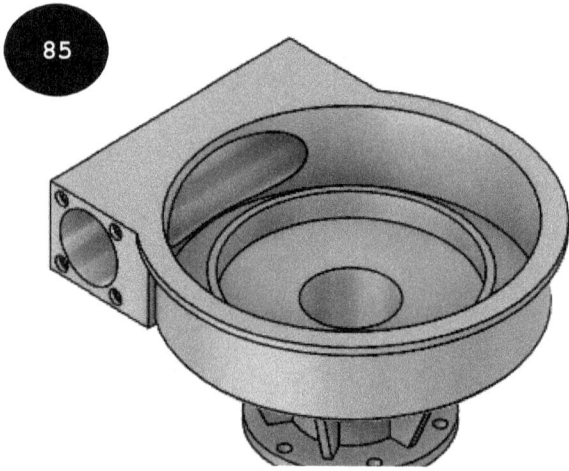

85

86

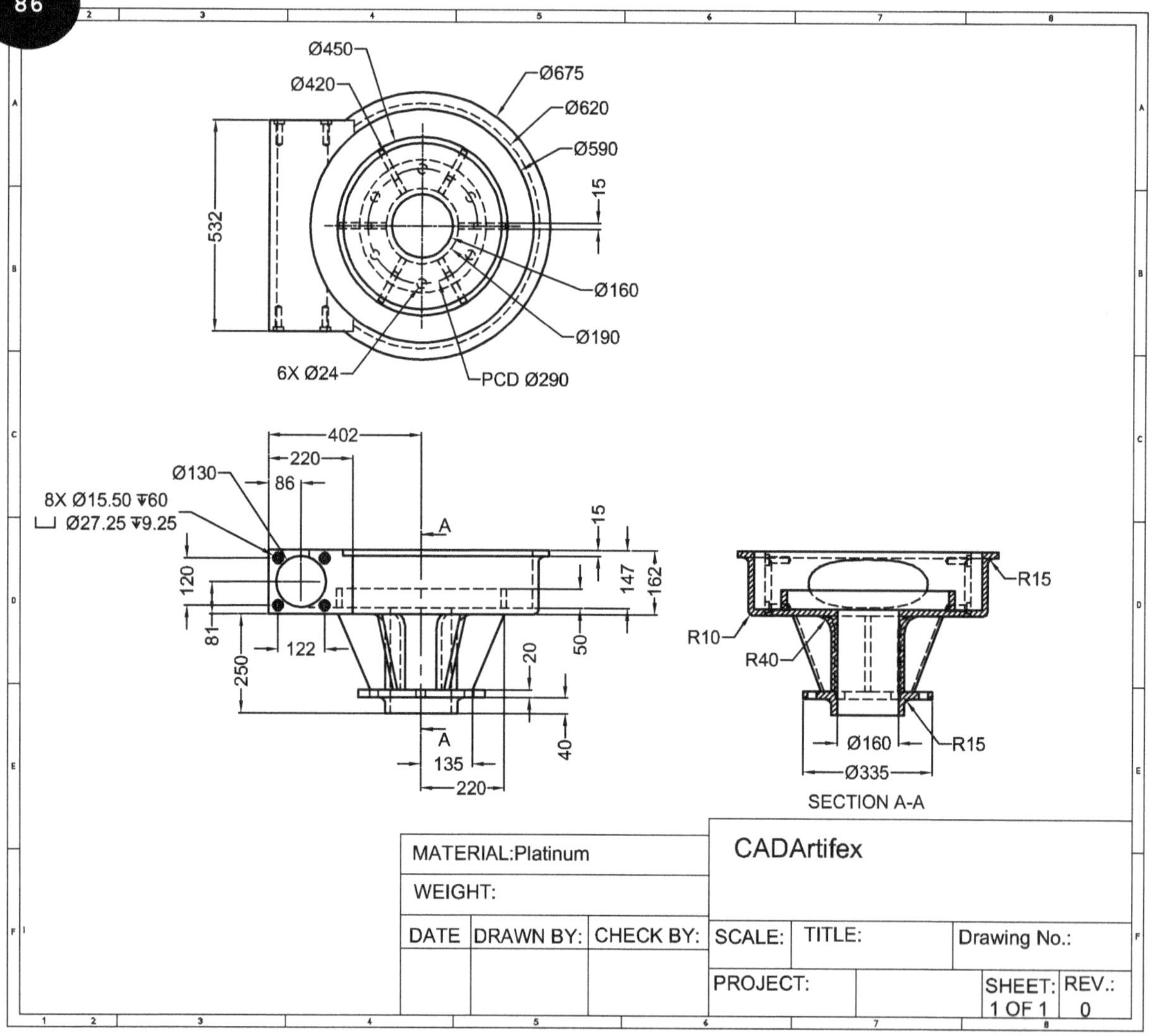

Ø450
Ø420
Ø675
Ø620
Ø590
15
532
Ø160
Ø190
6X Ø24
PCD Ø290

402
220
86
Ø130
8X Ø15.50 ⩒60
⨆ Ø27.25 ⩒9.25
A
15
147
162
120
81
250
122
20
50
A
135
40
220

R15
R10
R40
Ø160
R15
Ø335
SECTION A-A

MATERIAL:Platinum		CADArtifex		
WEIGHT:				
DATE	DRAWN BY:	CHECK BY:	SCALE:	TITLE:	Drawing No.:
			PROJECT:		SHEET: 1 OF 1	REV.: 0

Exercise 44.

Create the 3D model, as shown in Figure 87. Different views of the model and dimensions are shown in Figure 88. After creating the model, assign the Stainless Steel AISI 304 material and calculate its mass properties. All dimensions are in mm.

87

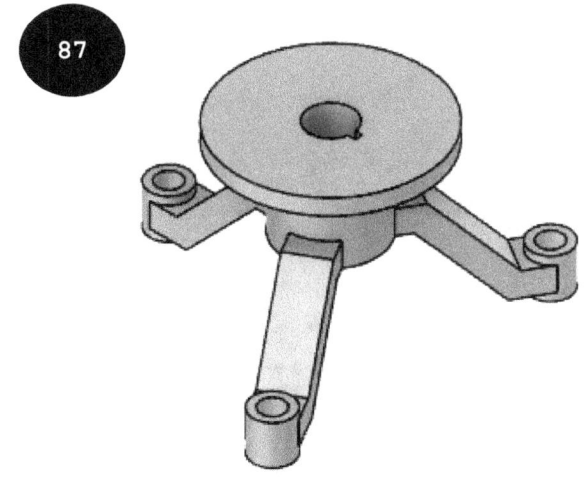

88

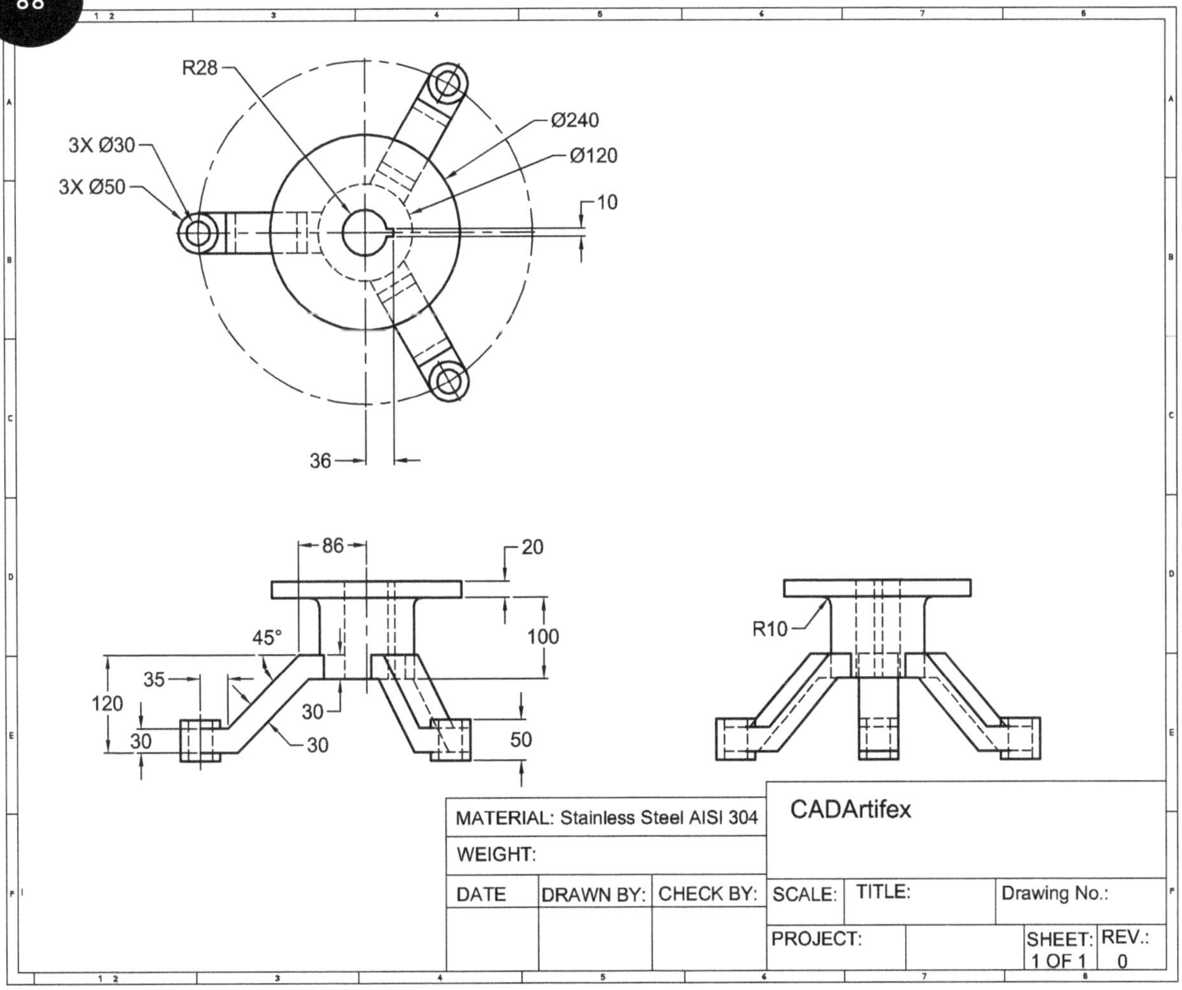

R28

3X Ø30

3X Ø50

Ø240

Ø120

10

36

86

20

100

45°

35

120

30

30

30

50

R10

MATERIAL: Stainless Steel AISI 304			CADArtifex		
WEIGHT:					
DATE	DRAWN BY:	CHECK BY:	SCALE:	TITLE:	Drawing No.:
			PROJECT:		SHEET: 1 OF 1 / REV.: 0

Exercise 45.

Create the 3D model, as shown in Figure 89. Different views of the model and dimensions are shown in Figure 90. After creating the model, assign the Steel AISI 1020 107 HR material and calculate its mass properties. All dimensions are in mm.

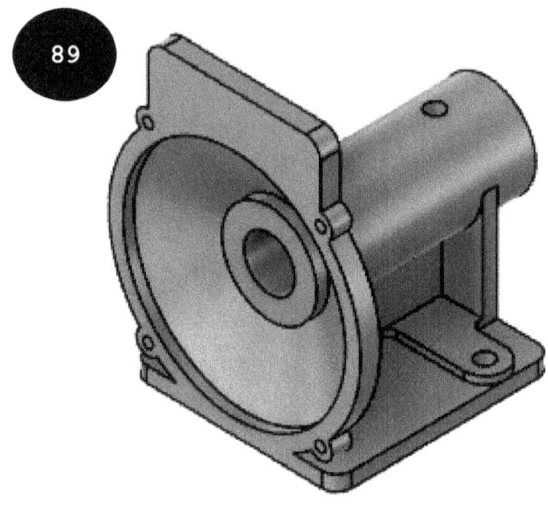

89

90

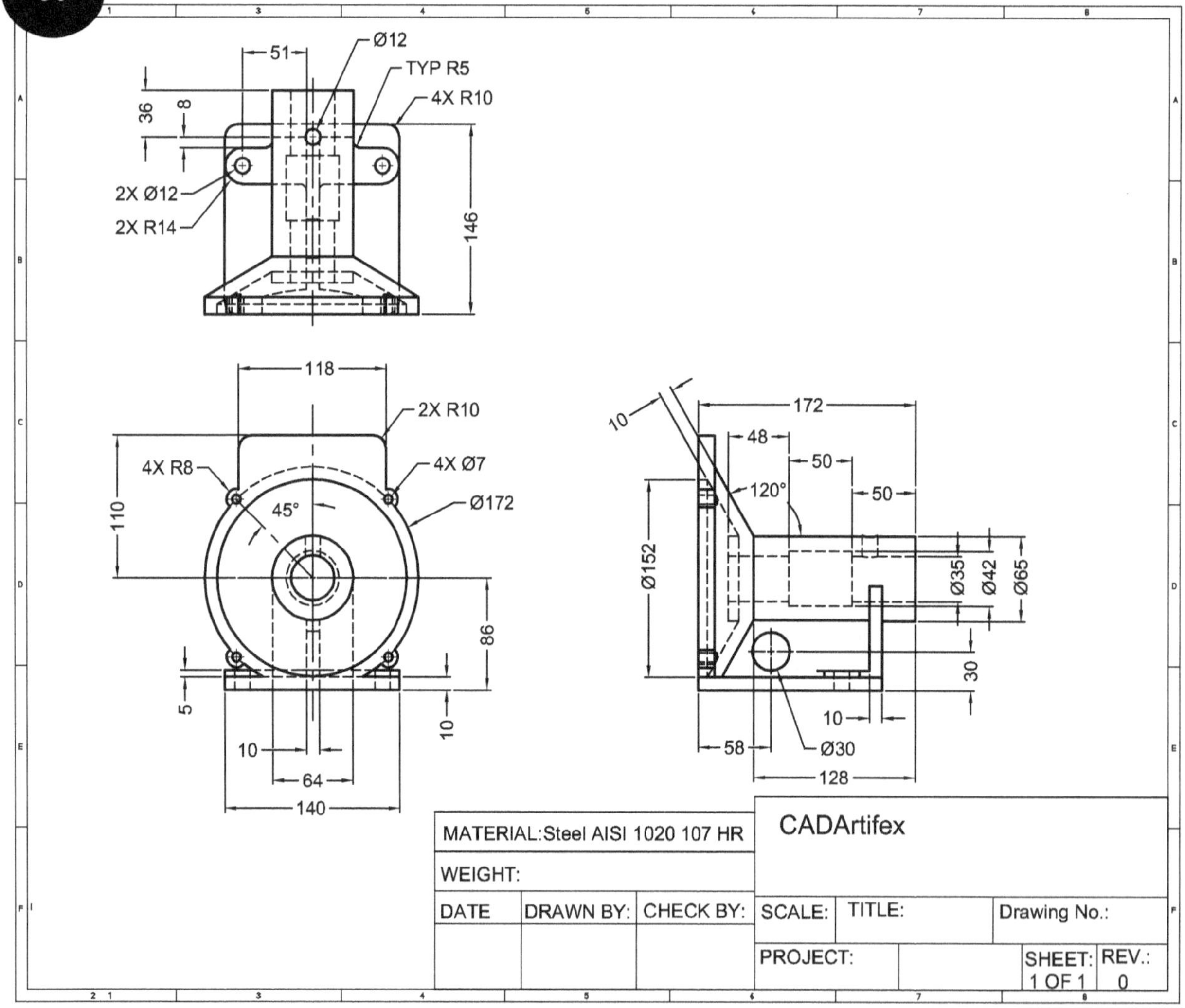

MATERIAL:Steel AISI 1020 107 HR			CADArtifex		
WEIGHT:					
DATE	DRAWN BY:	CHECK BY:	SCALE:	TITLE:	Drawing No.:
			PROJECT:		SHEET: 1 OF 1 / REV.: 0

Exercise 46.

Create the 3D model, as shown in Figure 91. Different views of the model and dimensions are shown in Figure 92. After creating the model, assign the Stainless Steel material and calculate its mass properties. All dimensions are in mm.

91

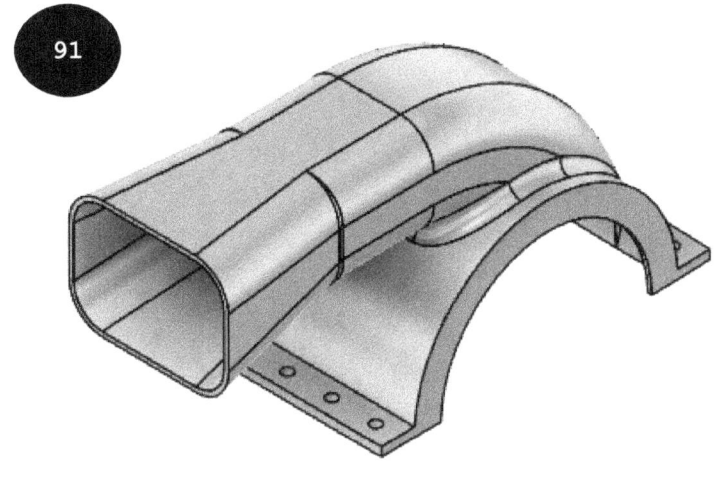

92

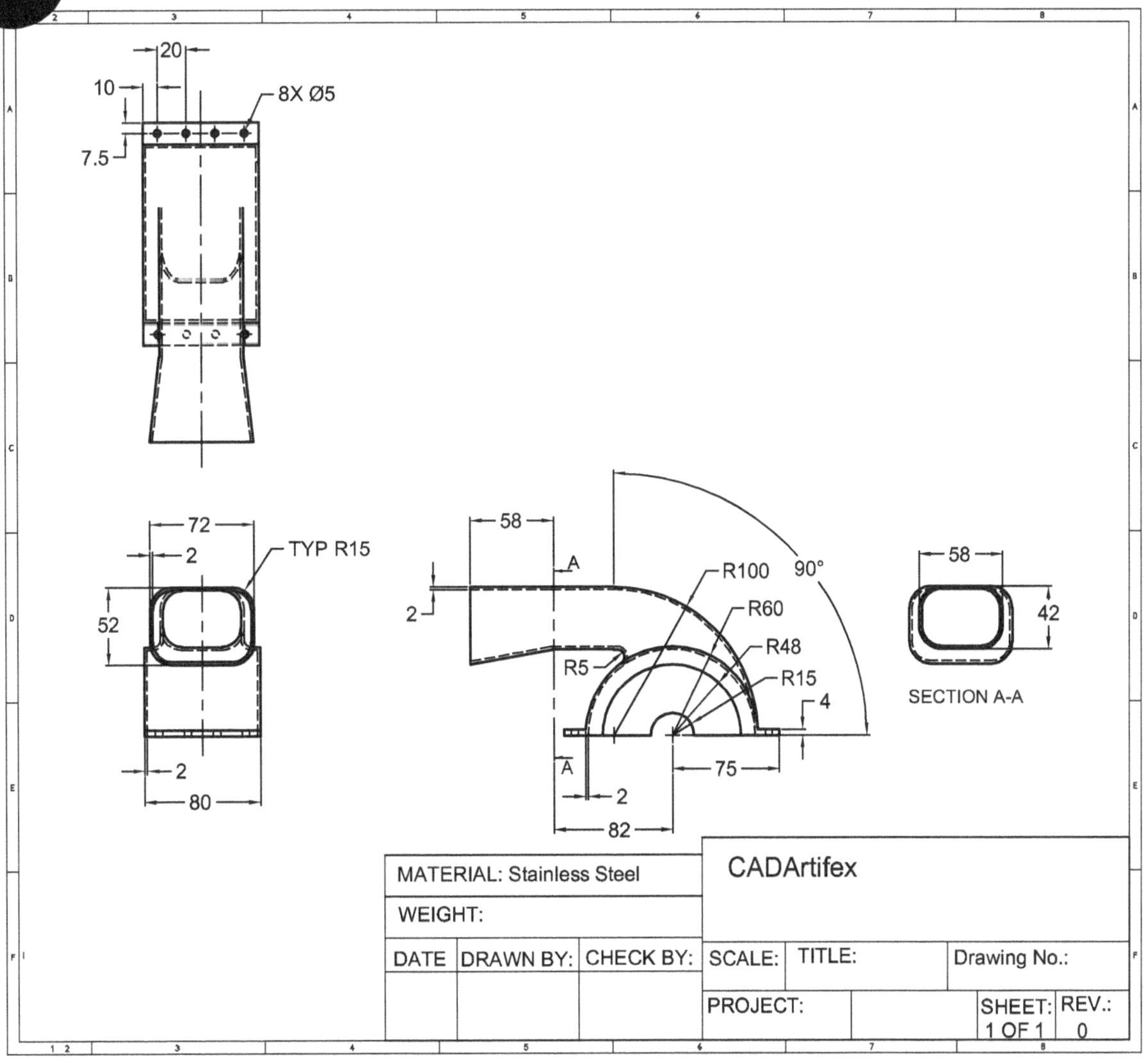

20
10
8X Ø5
7.5
72
2
TYP R15
52
2
80
58
A
2
R100
90°
R60
R48
R5
R15
4
A
75
2
82
58
42
SECTION A-A
MATERIAL: Stainless Steel
CADArtifex
WEIGHT:
DATE
DRAWN BY:
CHECK BY:
SCALE:
TITLE:
Drawing No.:
PROJECT:
SHEET:
1 OF 1
REV.:
0

Exercise 47.

Create the 3D model, as shown in Figure 93. Different views of the model and dimensions are shown in Figure 94. After creating the model, assign the Platinum material and calculate its mass properties. All dimensions are in mm.

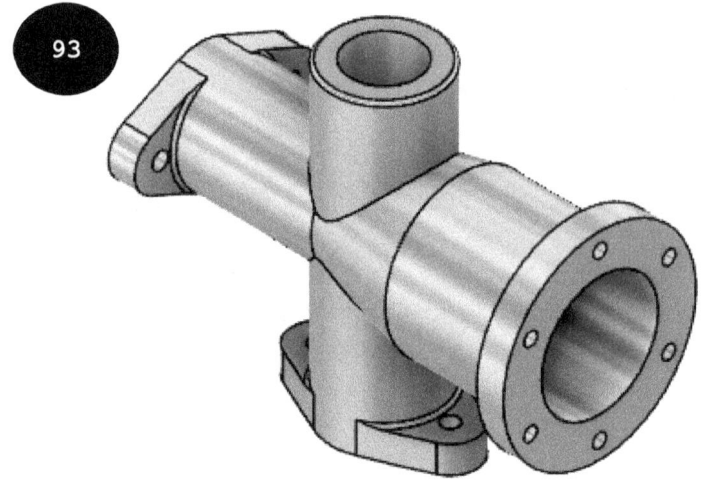

93

94

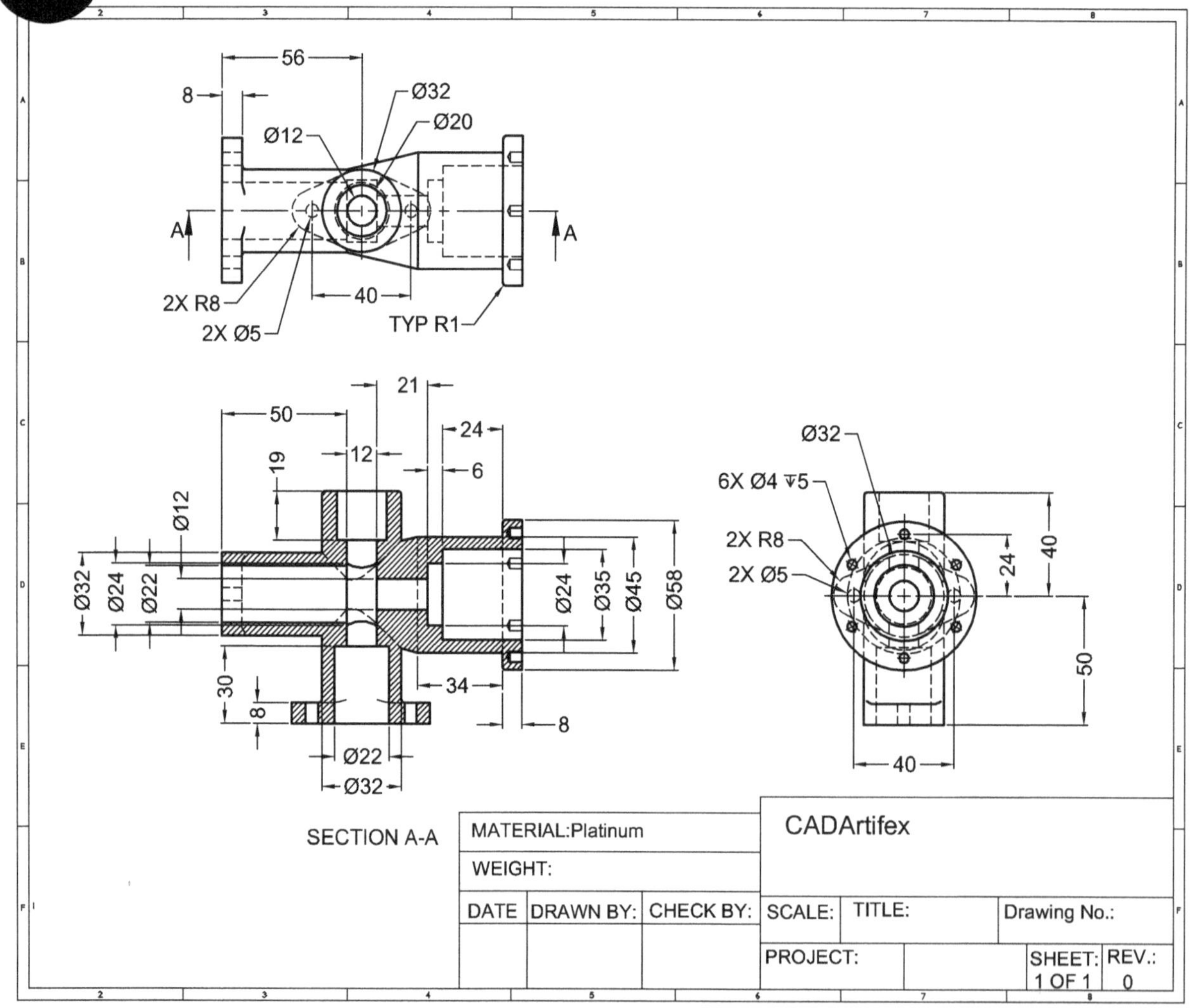

56
8
Ø12
Ø32
Ø20
A
A
2X R8
2X Ø5
40
TYP R1
21
50
24
12
6
19
Ø12
Ø32
Ø24
Ø22
Ø24
Ø35
Ø45
Ø58
30
8
34
8
Ø22
Ø32
SECTION A-A
Ø32
6X Ø4 ↧5
2X R8
2X Ø5
24
40
50
40
MATERIAL:Platinum
WEIGHT:
CADArtifex
DATE
DRAWN BY:
CHECK BY:
SCALE:
TITLE:
Drawing No.:
PROJECT:
SHEET:
1 OF 1
REV.:
0

Exercise 48.

Create the 3D model, as shown in Figure 95. Different views of the model and dimensions are shown in Figure 96. After creating the model, assign the Stainless Steel AISI 304 material and calculate its mass properties. All dimensions are in mm.

95

96

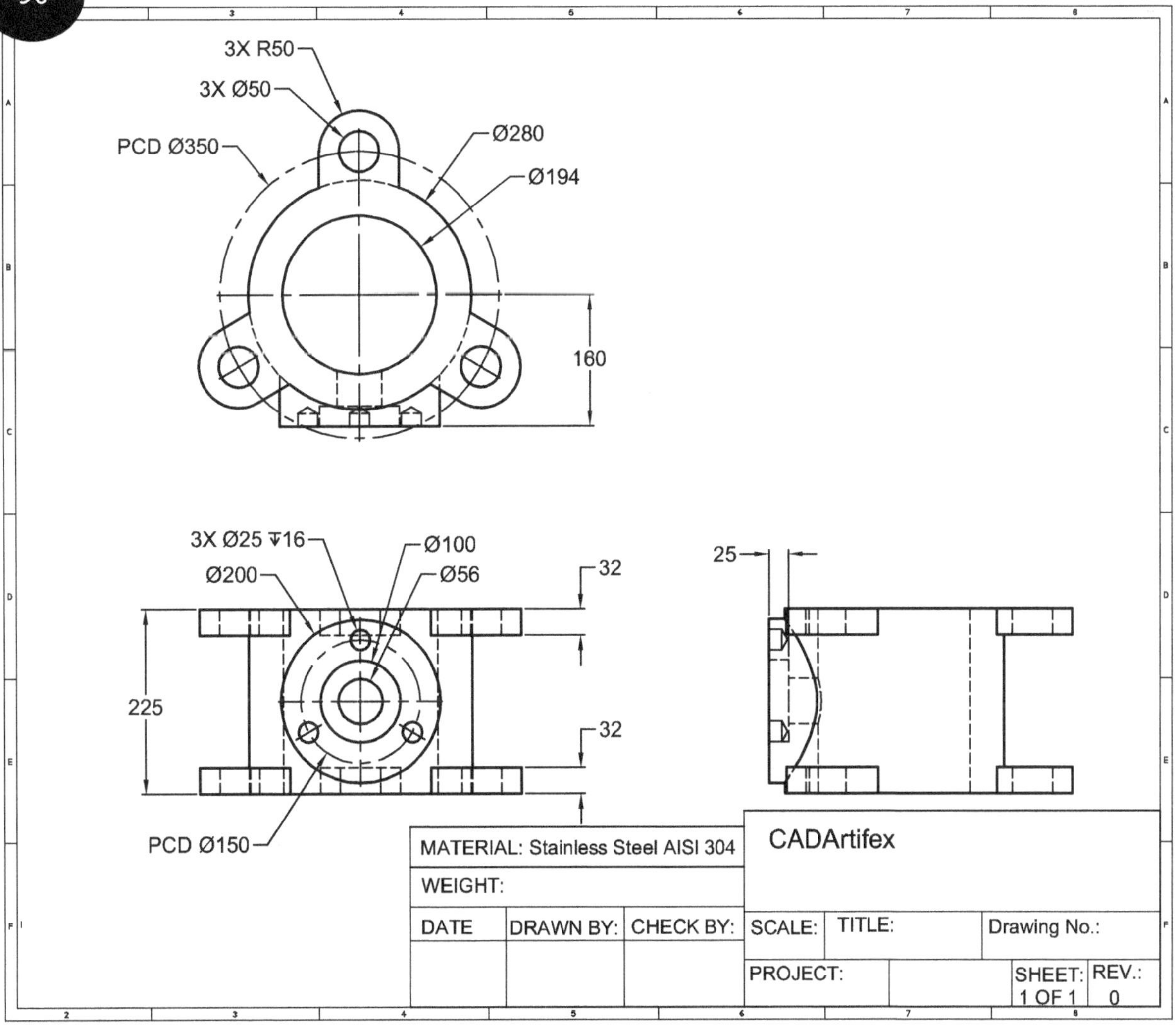

3X R50

3X Ø50

PCD Ø350

Ø280

Ø194

160

3X Ø25 ↧16

Ø200

Ø100

Ø56

32

25

Ø200

225

32

PCD Ø150

MATERIAL: Stainless Steel AISI 304	CADArtifex			
WEIGHT:				
DATE	DRAWN BY:	CHECK BY:	SCALE:	TITLE:	Drawing No.:
			PROJECT:		SHEET: 1 OF 1	REV.: 0

Exercise 49.

Create the 3D model, as shown in Figure 97. Different views of the model and dimensions are shown in Figure 98. After creating the model, assign the Steel AISI 1020 107 HR material and calculate its mass properties. All dimensions are in mm.

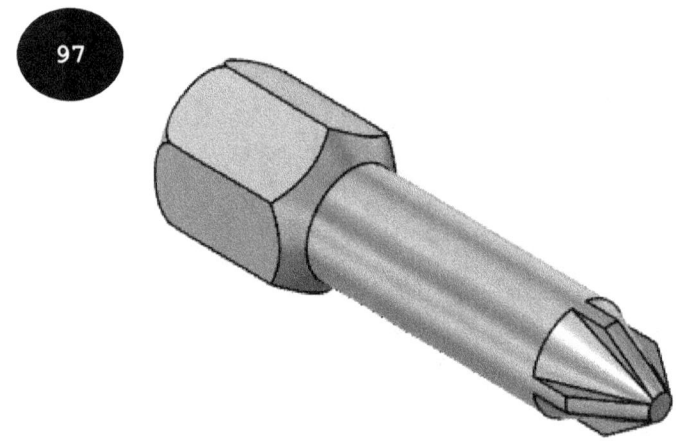

97

98

R5

68

35°

Ø30
Ø20
Ø5

135°

18

3

3

100

MATERIAL:Steel AISI 1020 107 HR			CADArtifex			
WEIGHT:						
DATE	DRAWN BY:	CHECK BY:	SCALE:	TITLE:		Drawing No.:
			PROJECT:			SHEET: 1 OF 1	REV.: 0

Exercise 50.

Create the 3D model, as shown in Figure 99. Different views of the model and dimensions are shown in Figure 100. After creating the model, assign the Stainless Steel material and calculate its mass properties. All dimensions are in mm.

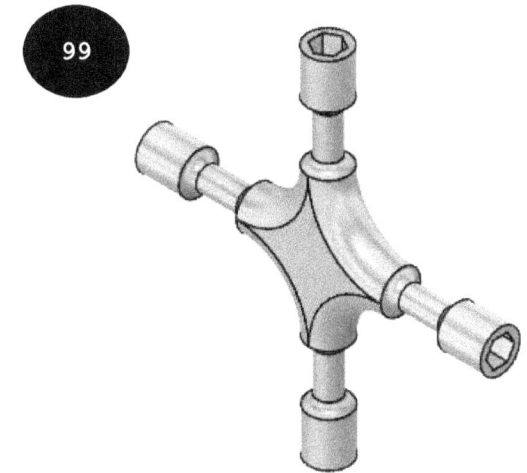

99

100

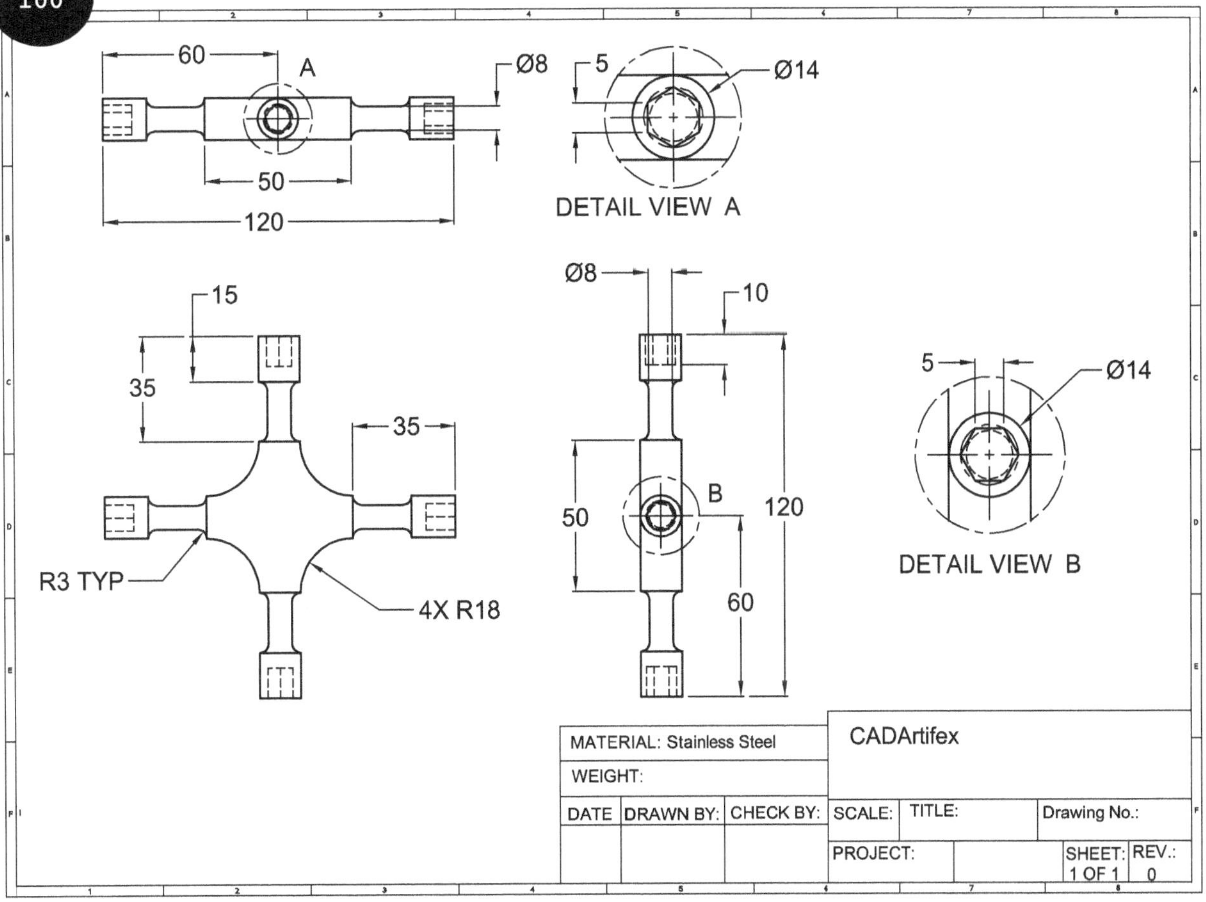

60
A
Ø8
5
Ø14
50
120
DETAIL VIEW A
15
35
35
R3 TYP
4X R18
Ø8
10
50
B
120
60
5
Ø14
DETAIL VIEW B
MATERIAL: Stainless Steel
WEIGHT:
DATE
DRAWN BY:
CHECK BY:
CADArtifex
SCALE:
TITLE:
Drawing No.:
PROJECT:
SHEET: 1 OF 1
REV.: 0

Exercise 51.

Create the 3D model, as shown in Figure 101. Different views of the model and dimensions are shown in Figure 102. After creating the model, assign the Stainless Steel material and calculate its mass properties. All dimensions are in mm.

101

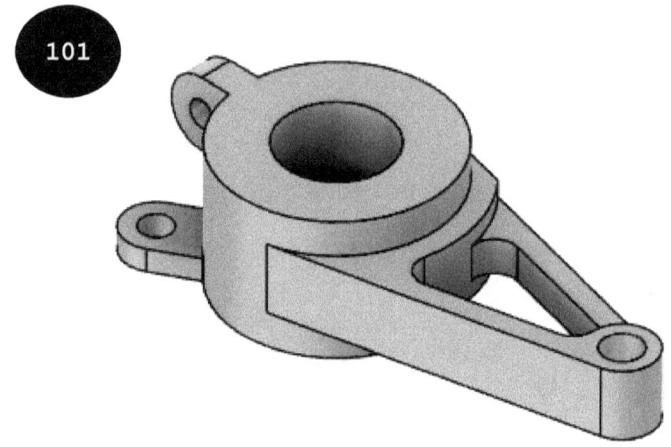

102

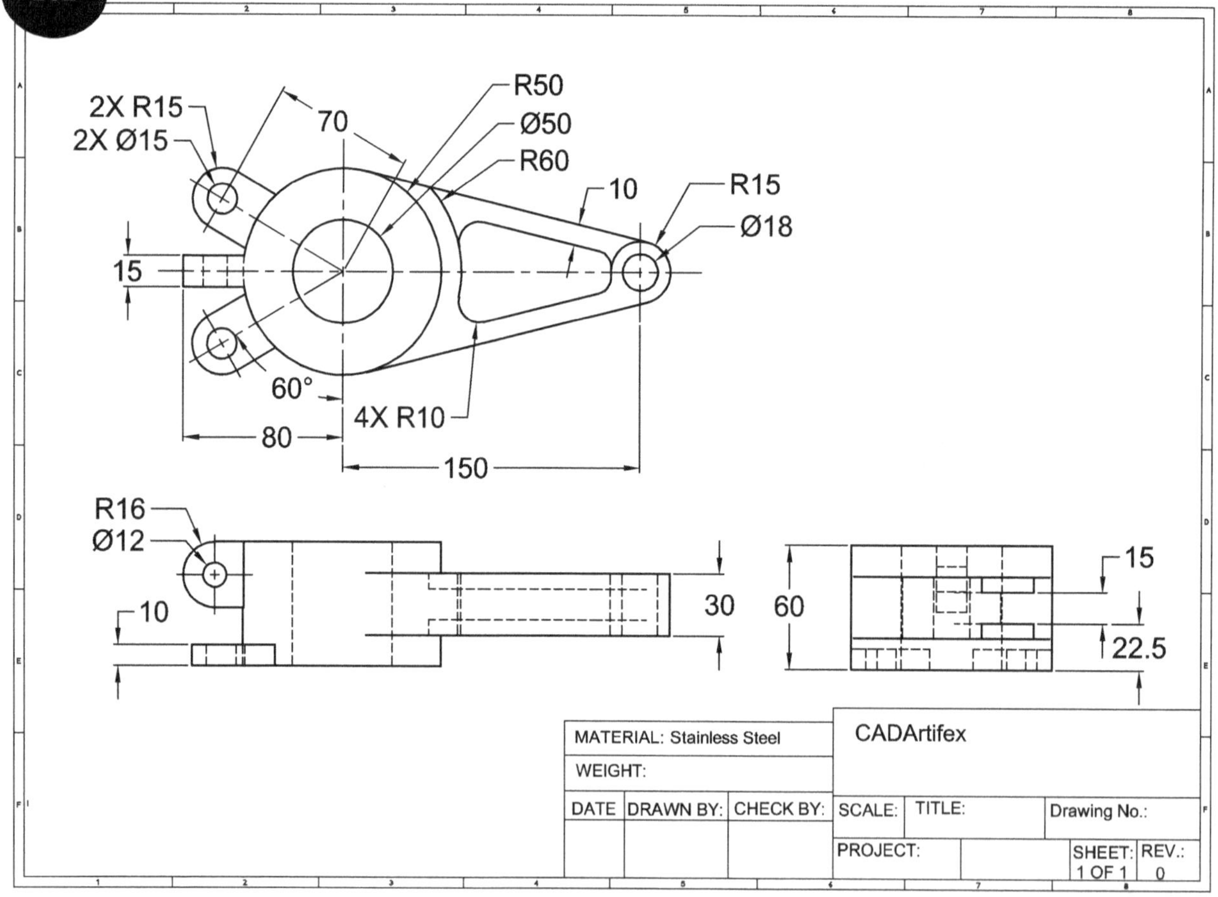

2X R15
2X Ø15
R50
Ø50
R60
70
10
R15
Ø18
15
60°
4X R10
80
150

R16
Ø12
10
30
60
15
22.5

MATERIAL: Stainless Steel		CADArtifex			
WEIGHT:					
DATE	DRAWN BY:	CHECK BY:	SCALE:	TITLE:	Drawing No.:
			PROJECT:		SHEET: REV.: 1 OF 1 0

Exercise 52.

Create the 3D model, as shown in Figure 103. Different views and dimensions are shown in Figure 104. After creating the model, assign the Platinum material and calculate its mass properties. All dimensions are in mm.

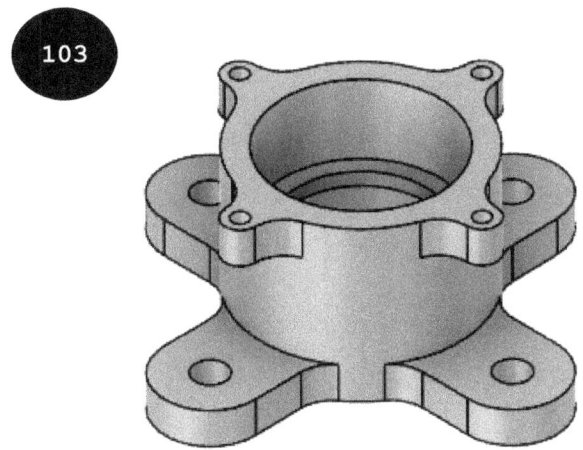

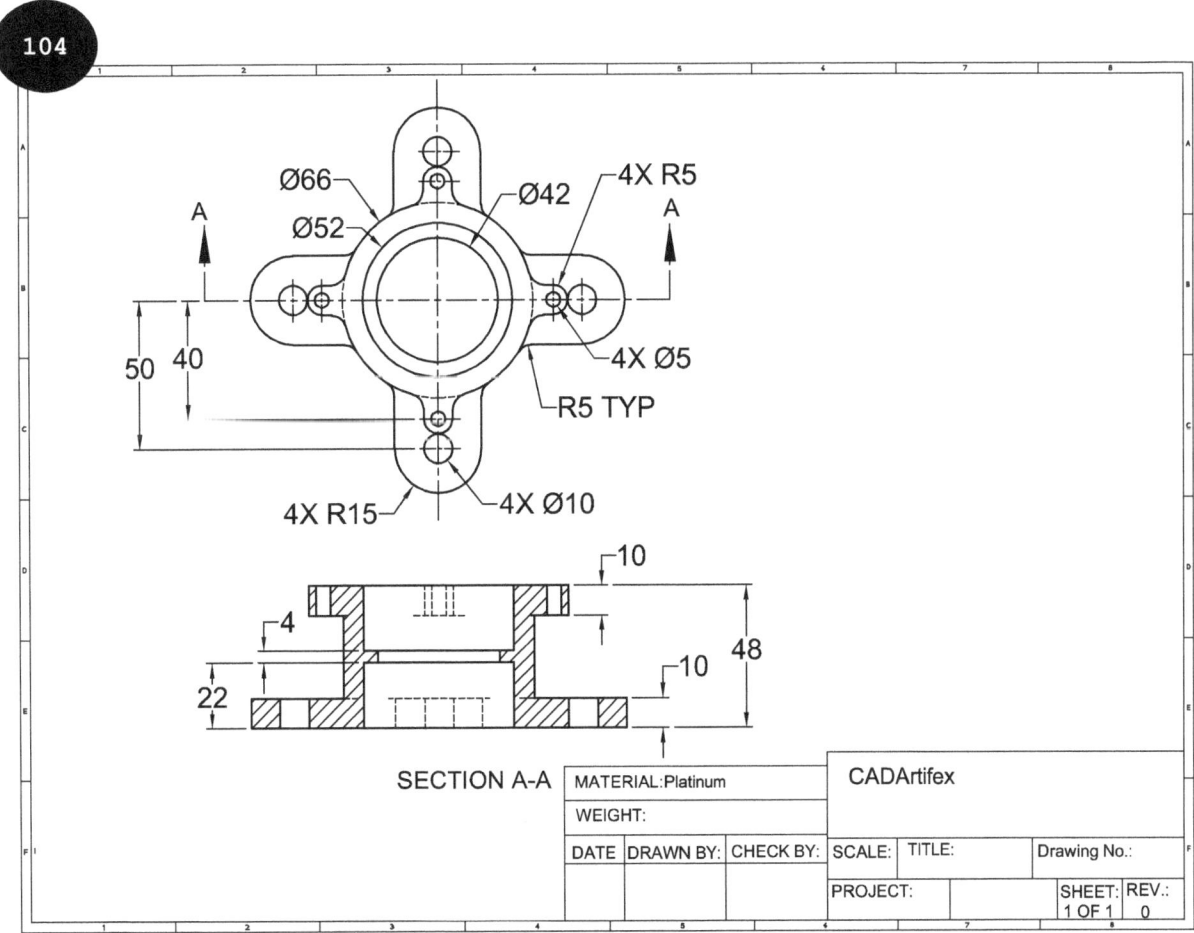

Exercise 53.

Create the 3D model, as shown in Figure 105. Different views of the model and dimensions are shown in Figure 106. After creating the model, assign the Platinum material and calculate its mass properties. All dimensions are in mm.

105

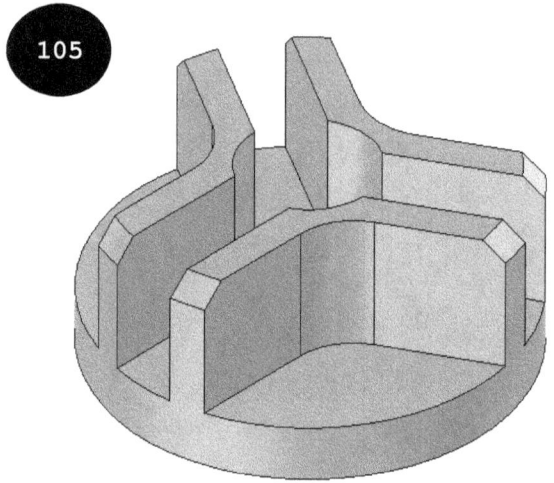

106

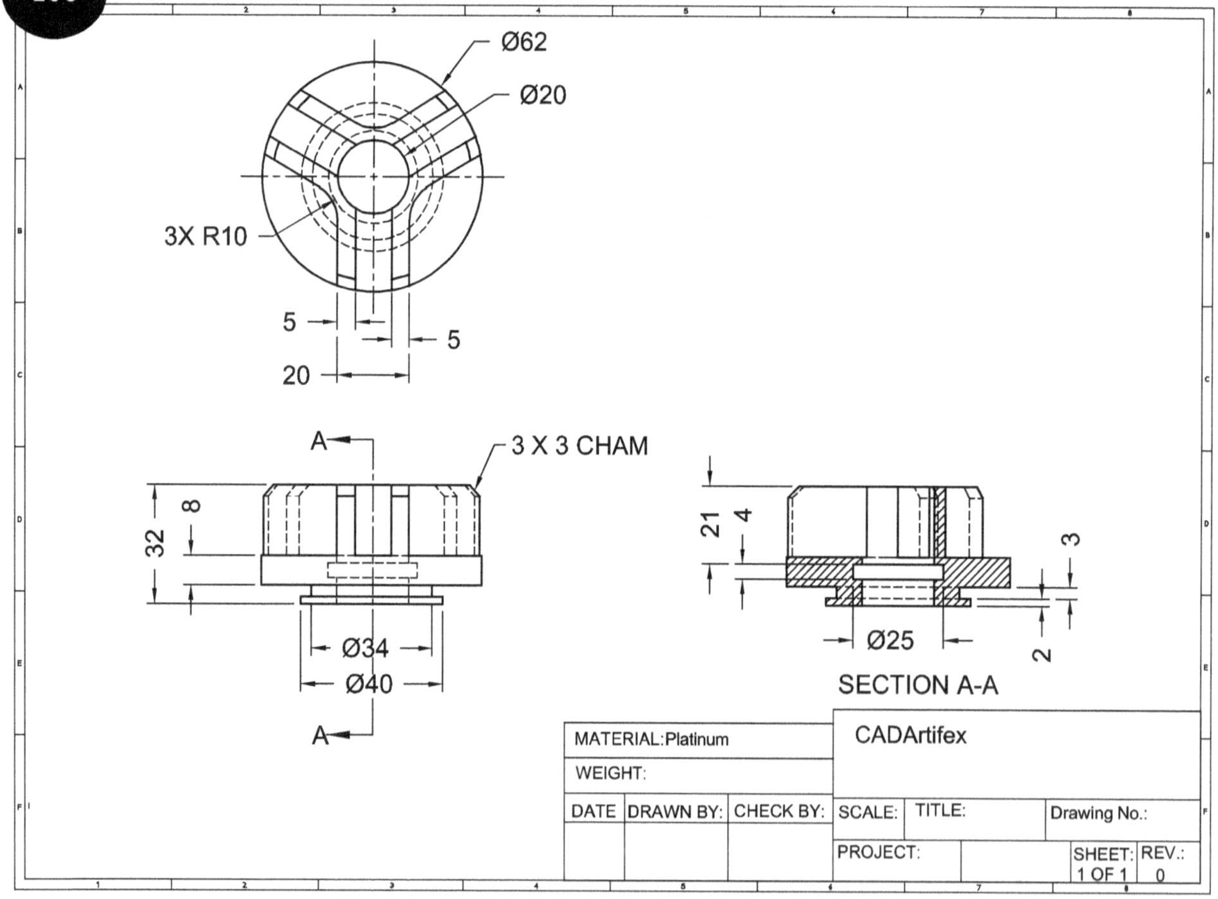

Ø62

Ø20

3X R10

5

5

20

A

3 X 3 CHAM

32

8

Ø34

Ø40

A

21 4

3

Ø25

2

SECTION A-A

MATERIAL:Platinum			CADArtifex		
WEIGHT:					
DATE	DRAWN BY:	CHECK BY:	SCALE:	TITLE:	Drawing No.:
			PROJECT:		SHEET: REV.: 1 OF 1 0

Exercise 54.

Create the 3D model, as shown in Figure 107. Different views of the model and dimensions are shown in Figure 108. After creating the model, assign the Stainless Steel AISI 304 material and calculate its mass properties. All dimensions are in mm.

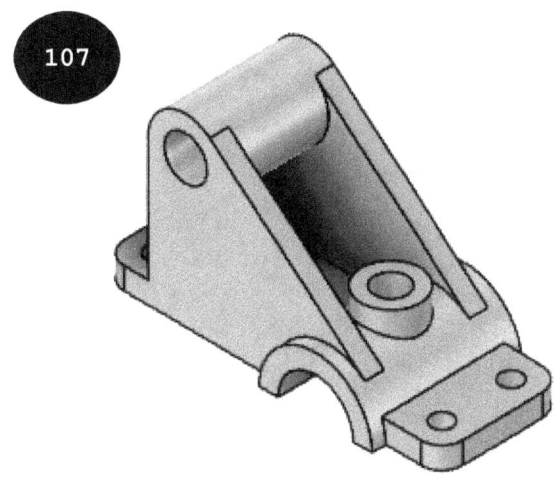

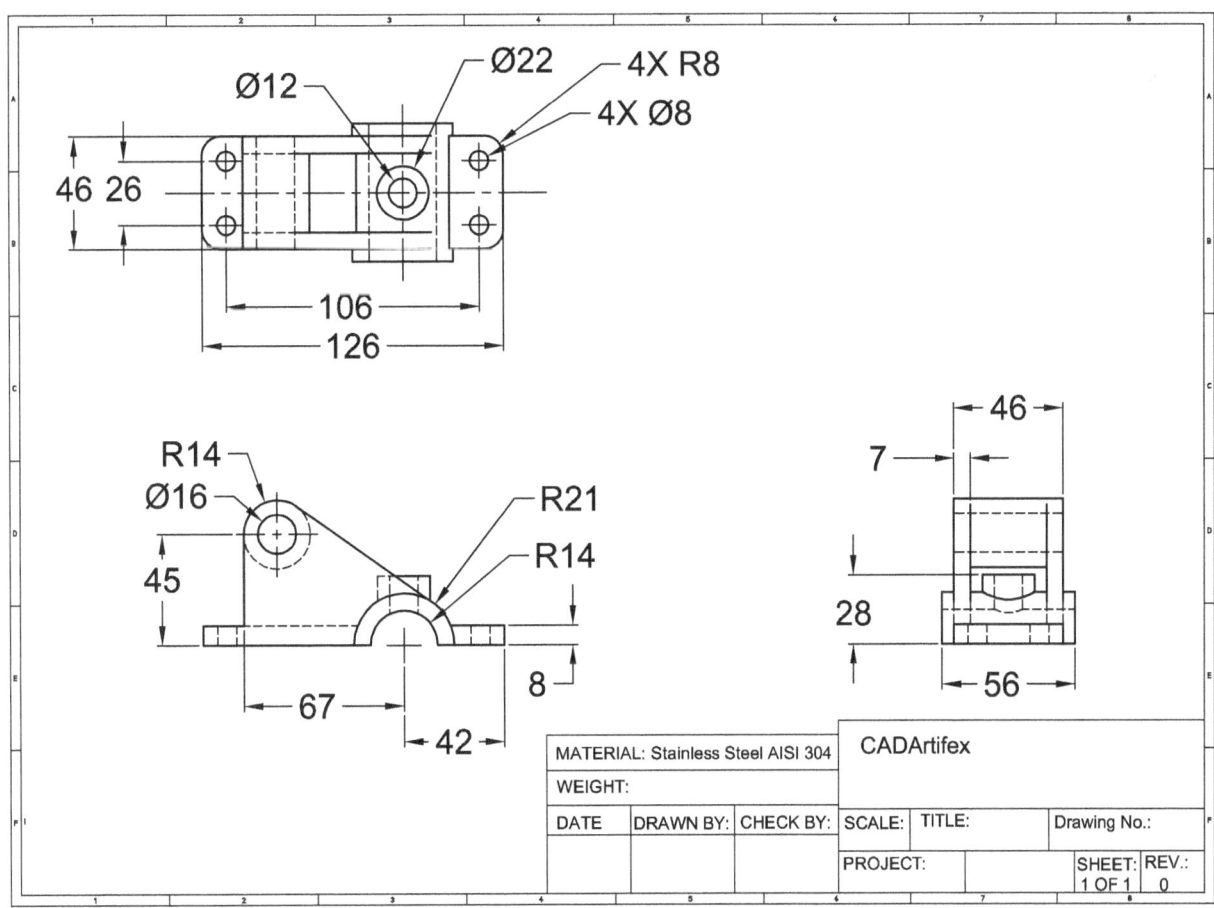

Exercise 55.

Create the 3D model, as shown in Figure 109. Different views of the model and dimensions are shown in Figure 110. After creating the model, assign the Stainless Steel AISI 304 material and calculate its mass properties. All dimensions are in mm.

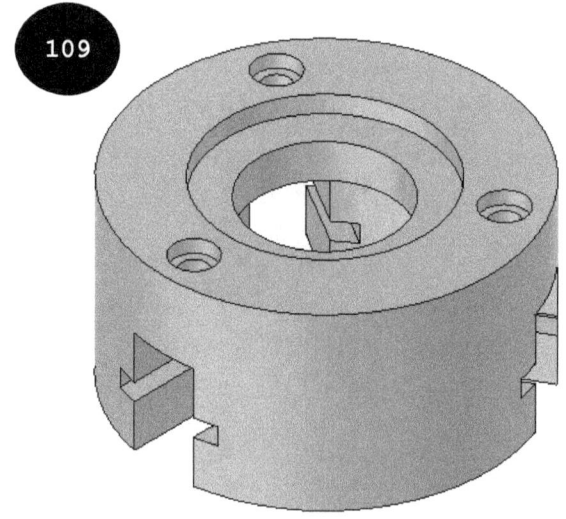

109

110

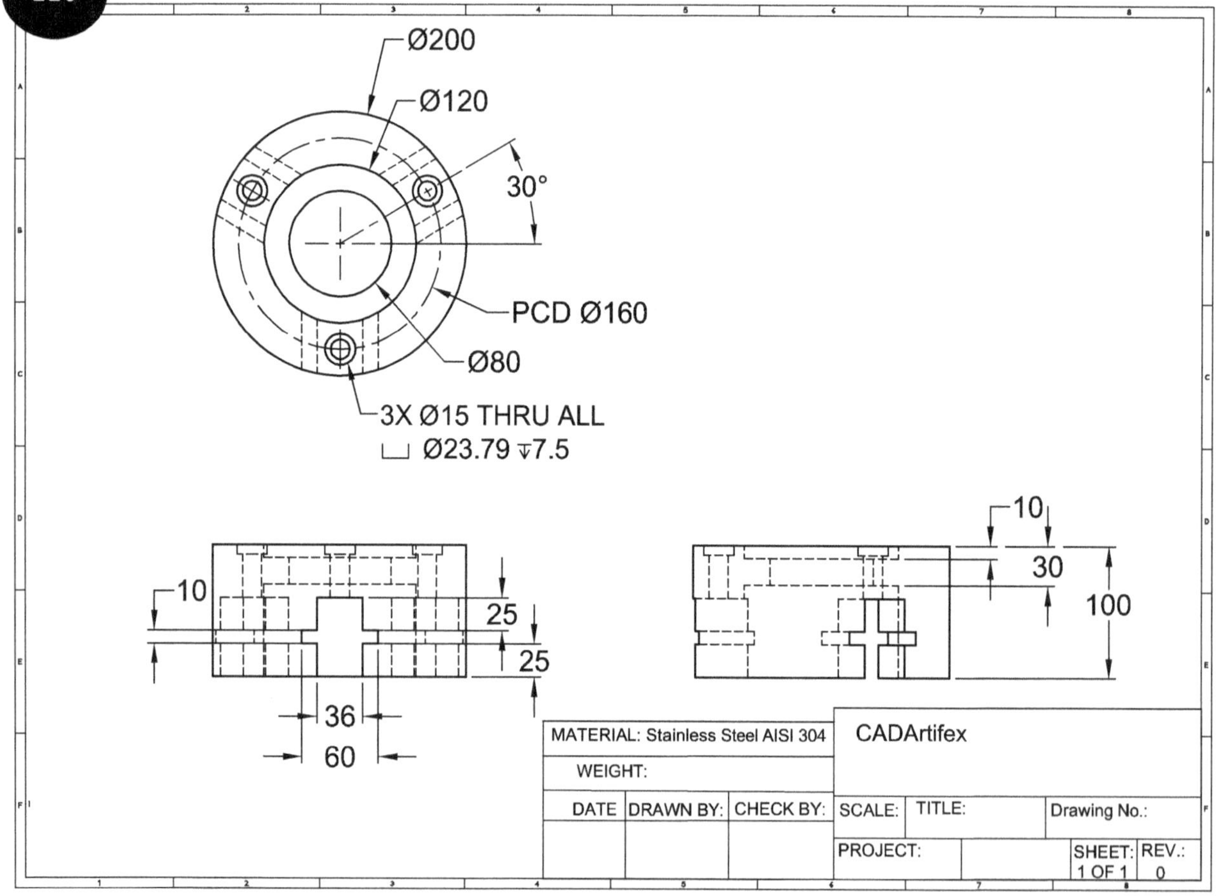

Ø200

Ø120

30°

PCD Ø160

Ø80

3X Ø15 THRU ALL
⌴ Ø23.79 ⍀7.5

10

10
30
100

10
25
25

36
60

MATERIAL: Stainless Steel AISI 304	CADArtifex			
WEIGHT:				
DATE	DRAWN BY:	CHECK BY:	SCALE:	TITLE:	Drawing No.:
			PROJECT:		SHEET: 1 OF 1	REV.: 0

Exercise 56.

Create the 3D model, as shown in Figure 111. Different views of the model and dimensions are shown in Figure 112. After creating the model, assign the Steel AISI 1020 107 HR material and calculate its mass properties. All dimensions are in mm.

111

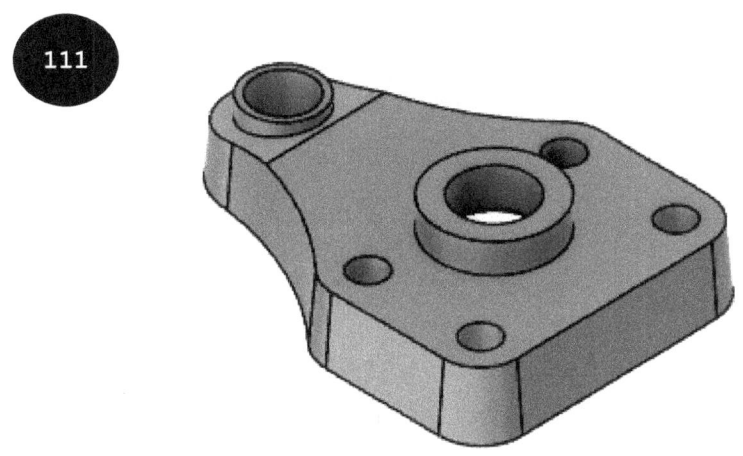

112

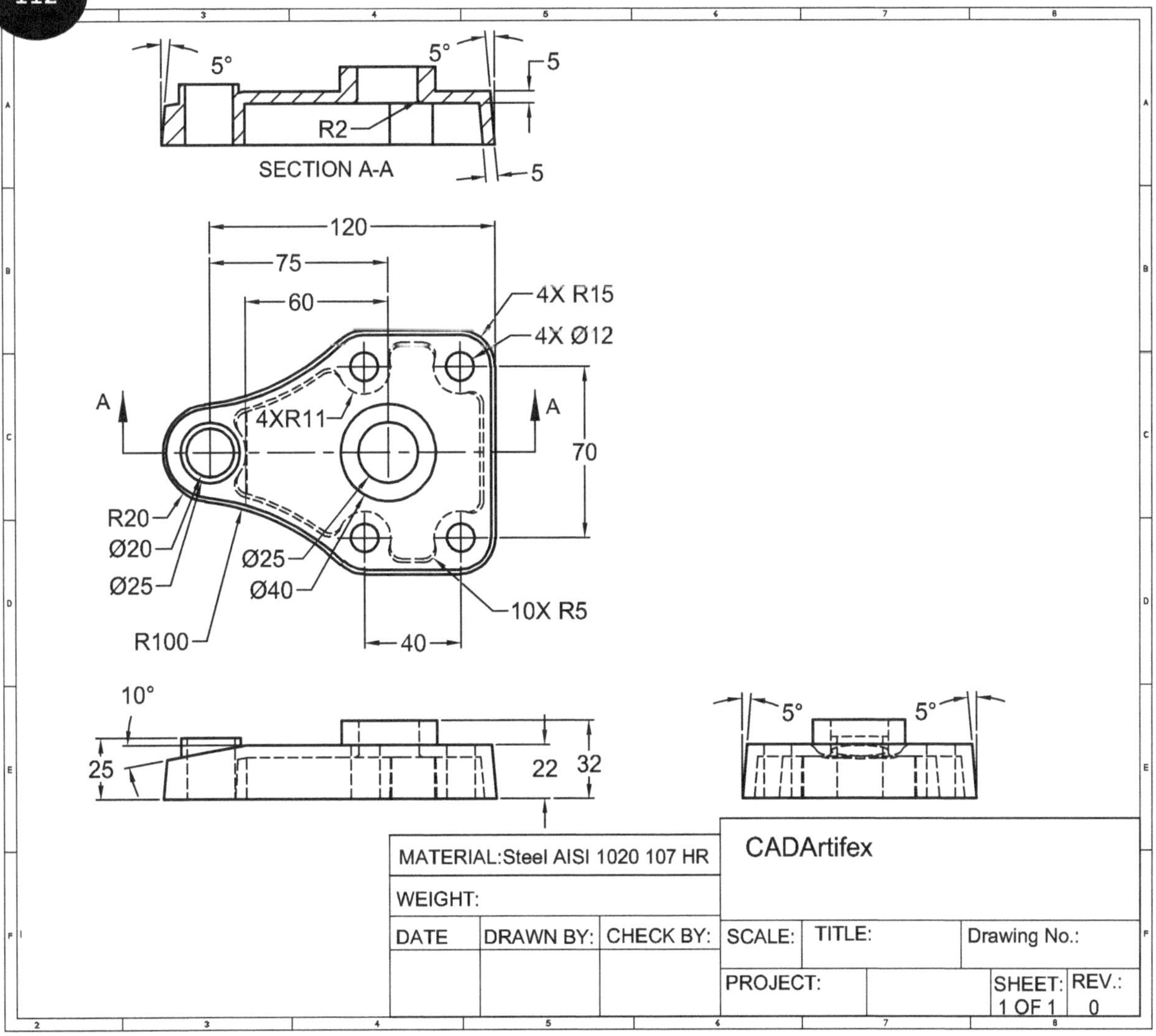

SECTION A-A

5°   5°   5

R2

5

120

75

60

4X R15

4X Ø12

A   4XR11   A

70

R20
Ø20   Ø25
Ø25   Ø40

10X R5

R100   40

10°

25   22 32

5°   5°

MATERIAL:Steel AISI 1020 107 HR		CADArtifex		
WEIGHT:				
DATE	DRAWN BY:	CHECK BY:	SCALE: TITLE:	Drawing No.:
			PROJECT:	SHEET: REV.: 1 OF 1 0

Exercise 57.

Create the 3D model, as shown in Figure 113. Different views and dimensions are shown in Figure 114. After creating the model, assign the Platinum material and calculate its mass properties. All dimensions are in mm.

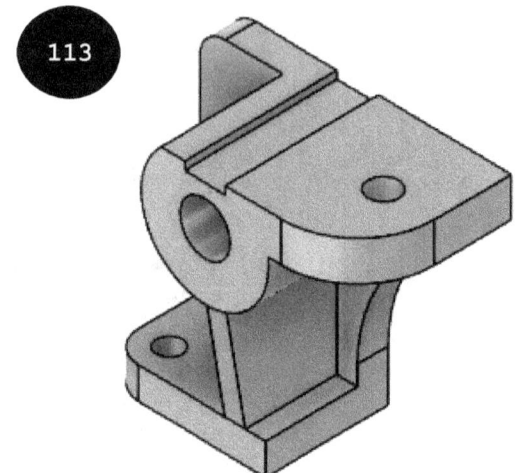

113

114

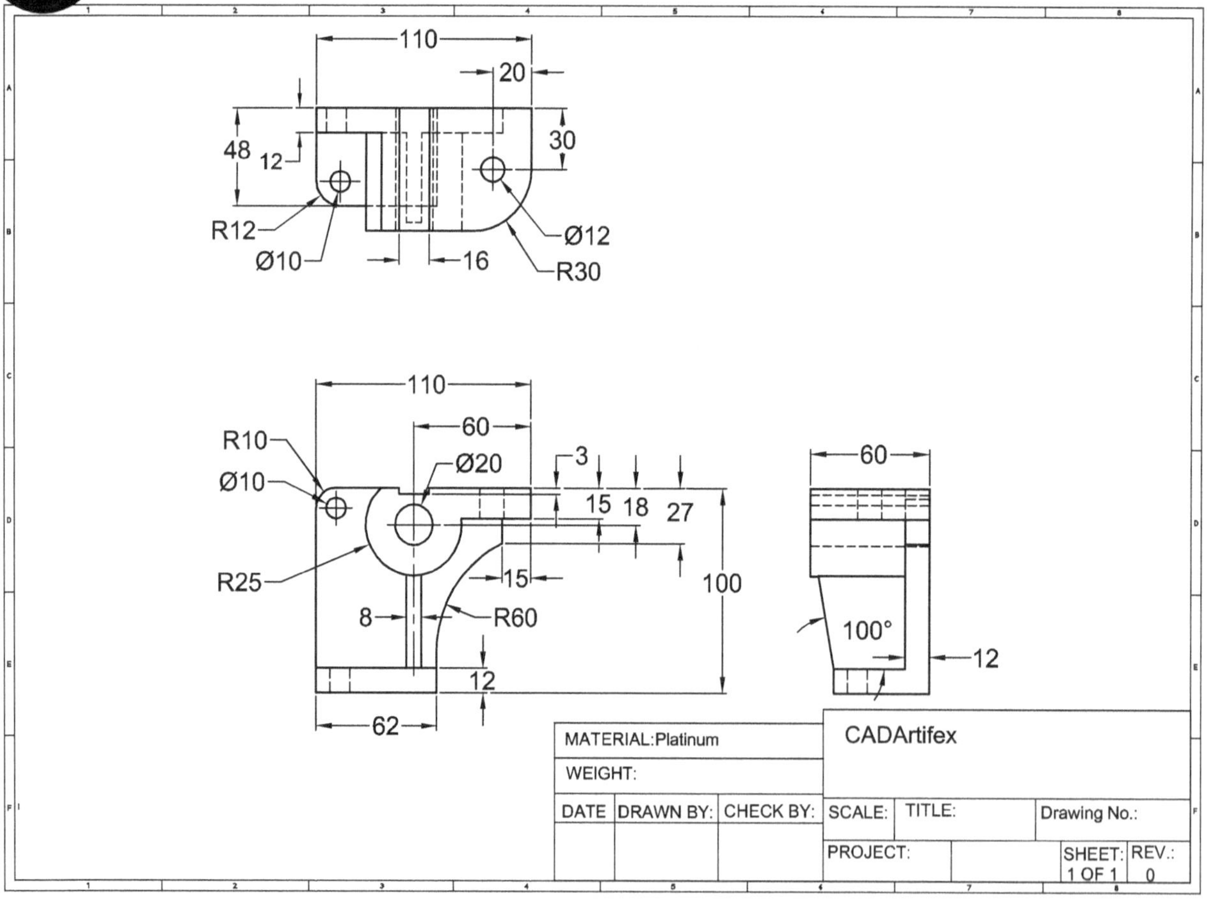

110
20
48
12
30
R12
Ø10
16
Ø12
R30
110
60
R10
Ø10
Ø20
3
15
18
27
R25
15
100
8
R60
12
62
60
100°
12
MATERIAL:Platinum
WEIGHT:
DATE
DRAWN BY:
CHECK BY:
CADArtifex
SCALE:
TITLE:
Drawing No.:
PROJECT:
SHEET:
1 OF 1
REV.:
0

Exercise 58.

Create the 3D model, as shown in Figure 115. Different views of the model and dimensions are shown in Figure 116. After creating the model, assign the Stainless Steel material and calculate its mass properties. All dimensions are in mm.

115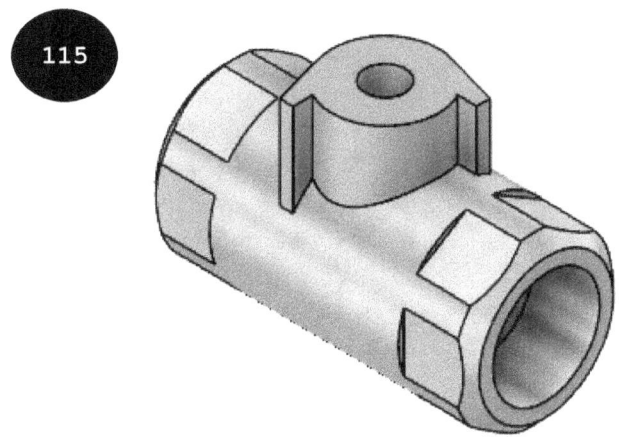

116

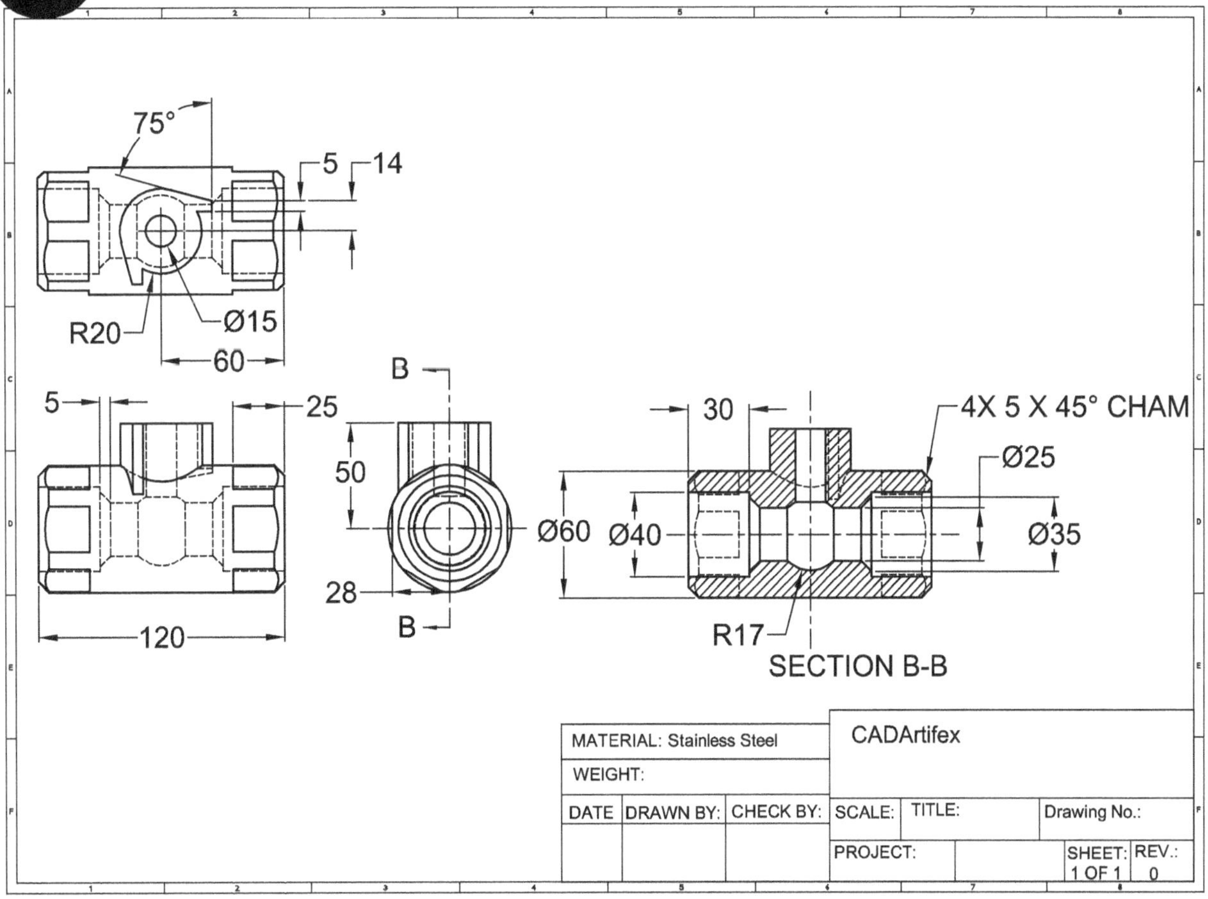

75°

5 ─14

R20 ─── Ø15

60

5 ─── 25

50

28

120

B

B

Ø60 Ø40

30

4X 5 X 45° CHAM

Ø25

Ø35

R17

SECTION B-B

MATERIAL: Stainless Steel	CADArtifex			
WEIGHT:				
DATE	DRAWN BY:	CHECK BY:	SCALE:	TITLE:	Drawing No.:
			PROJECT:		SHEET: 1 OF 1	REV.: 0

Exercise 59.

Create the 3D model, as shown in Figure 117. Different views of the model and dimensions are shown in Figure 118. After creating the model, assign the Platinum material and calculate its mass properties. All dimensions are in mm.

117

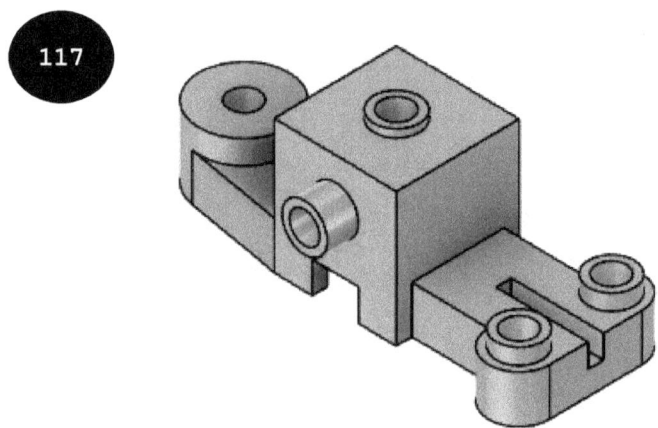

118

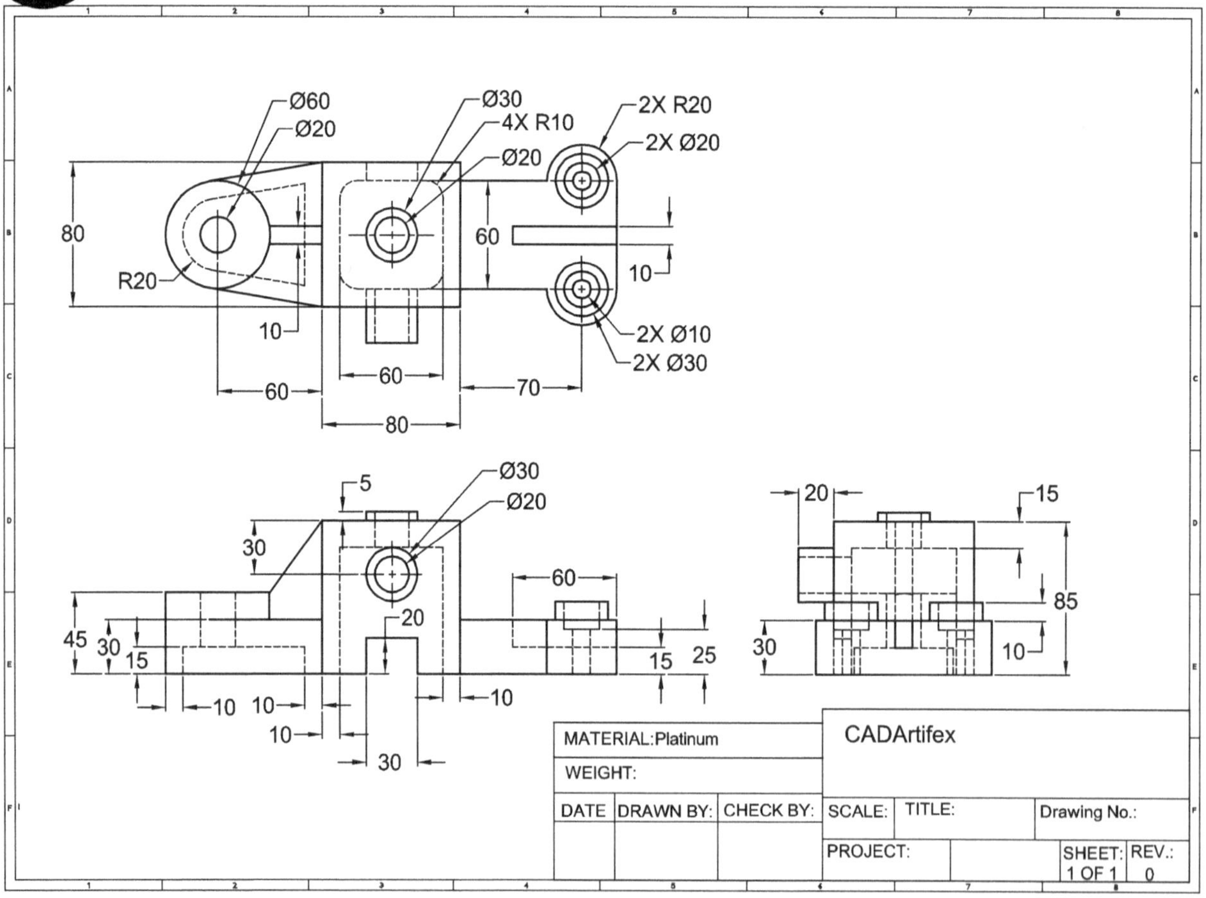

Exercise 60.

Create the 3D model, as shown in Figure 119. Different views of the model and dimensions are shown in Figure 120. After creating the model, assign the Stainless Steel AISI 304 material and calculate its mass properties. All dimensions are in mm.

119

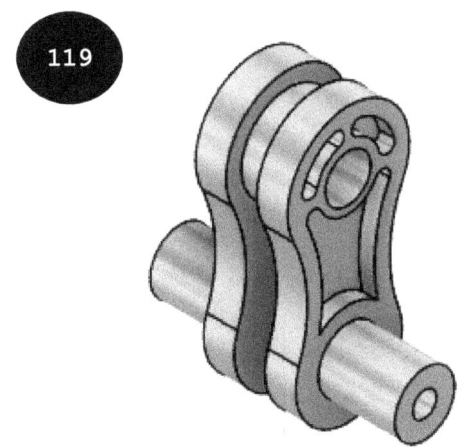

120

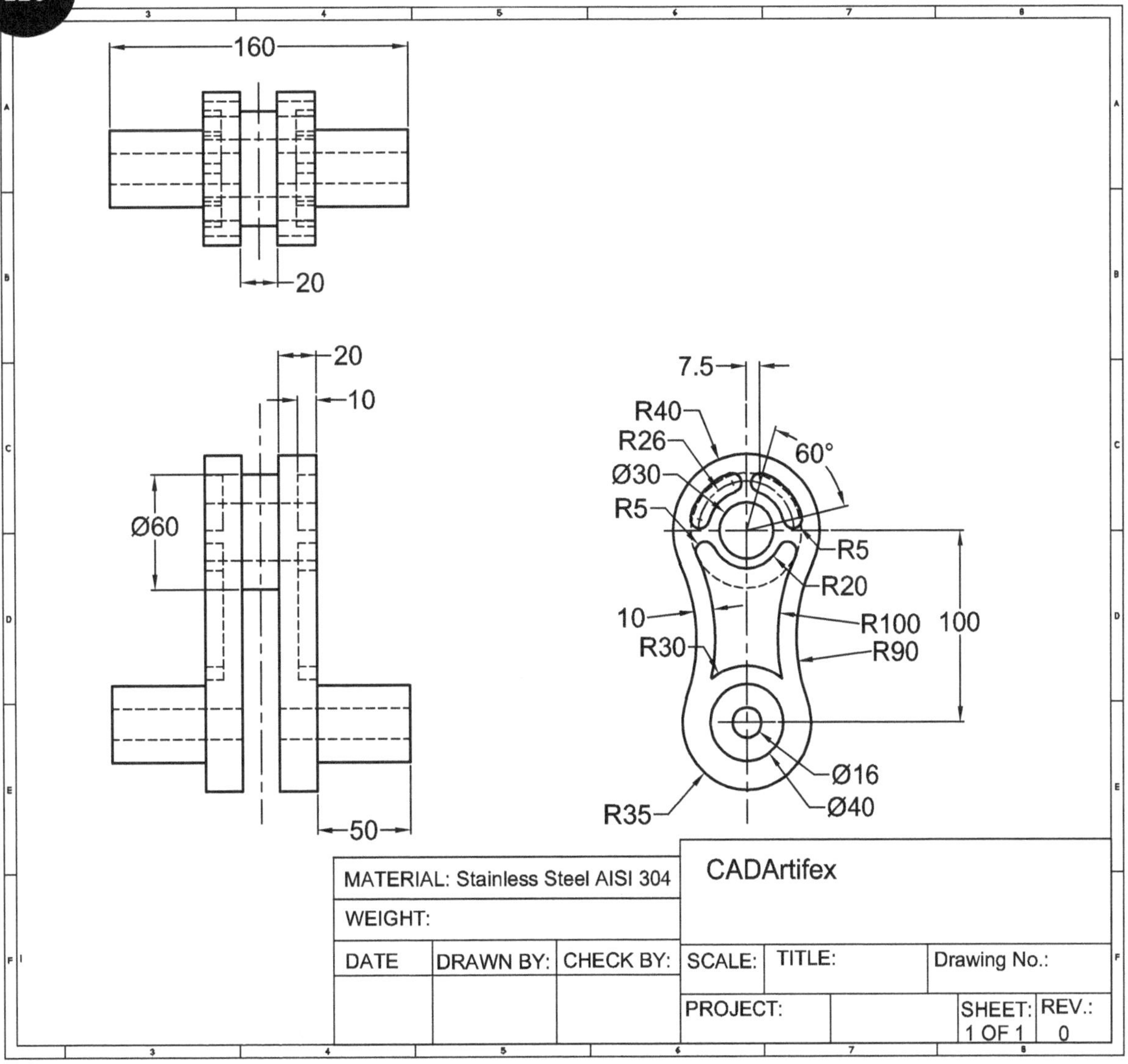

MATERIAL: Stainless Steel AISI 304

WEIGHT:

CADArtifex

DATE	DRAWN BY:	CHECK BY:	SCALE:	TITLE:		Drawing No.:
			PROJECT:			SHEET: 1 OF 1	REV.: 0

Dimensions shown: 160, 20, 20, 10, Ø60, 50, 7.5, R40, R26, Ø30, R5, 60°, R5, R20, 10, R100, 100, R30, R90, Ø16, Ø40, R35

Exercise 61.

Create the 3D model, as shown in Figure 121. Different views of the model and dimensions are shown in Figure 122. After creating the model, assign the Stainless Steel AISI 304 material and calculate its mass properties. All dimensions are in mm.

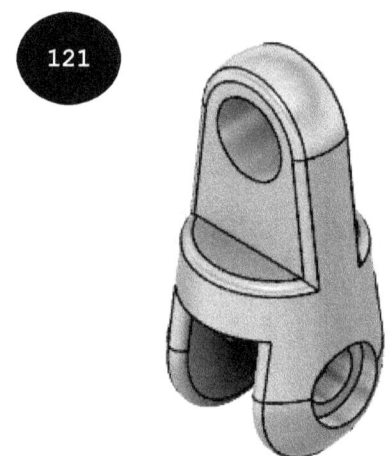

121

122

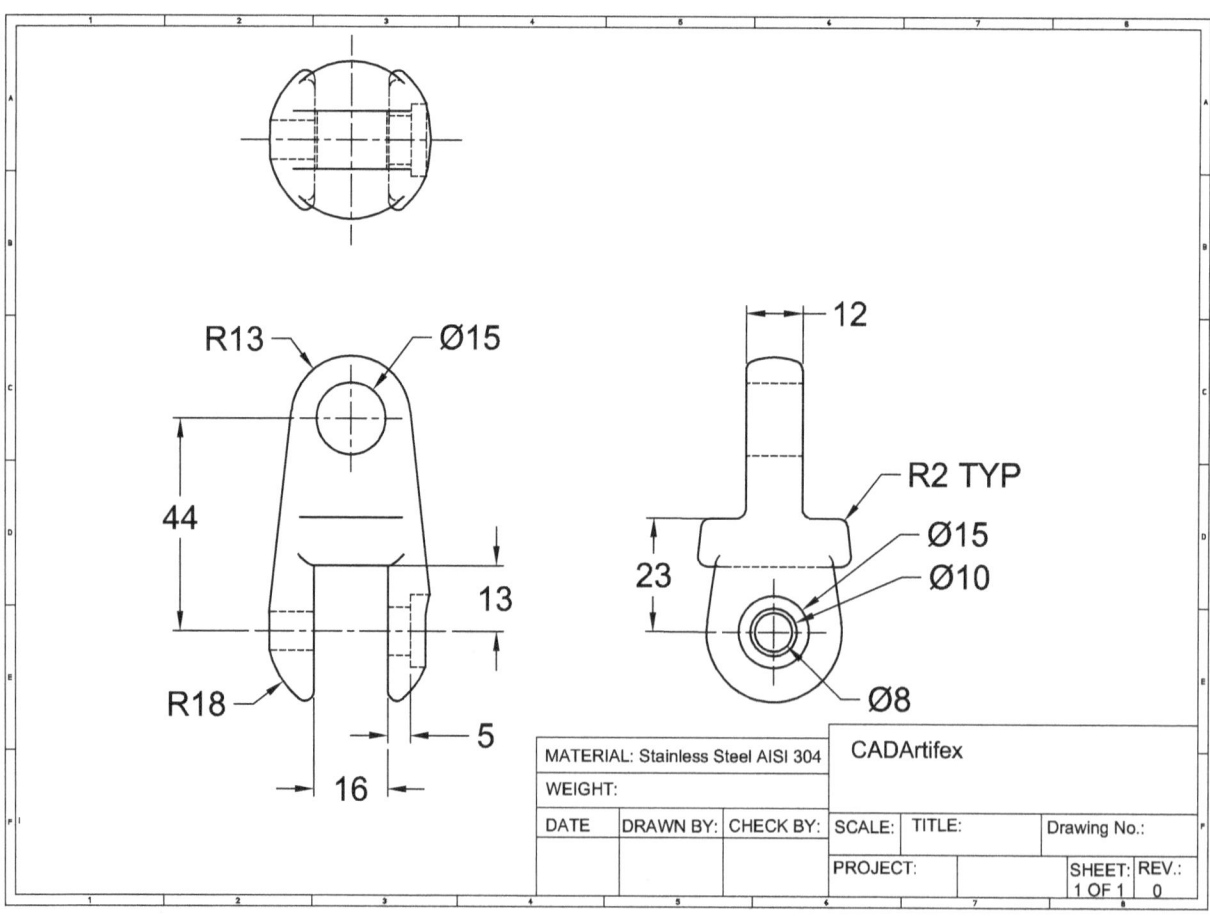

Exercise 62.

Create the 3D model, as shown in Figure 123. Different views of the model and dimensions are shown in Figure 124. After creating the model, assign the Platinum material and calculate its mass properties. All dimensions are in mm.

123

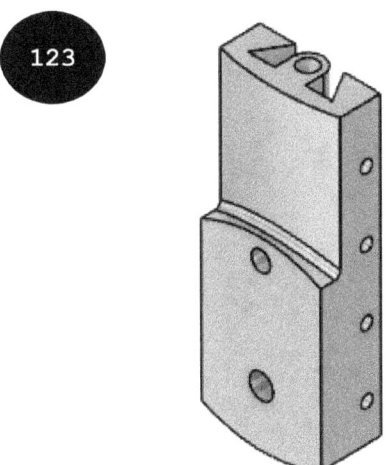

124

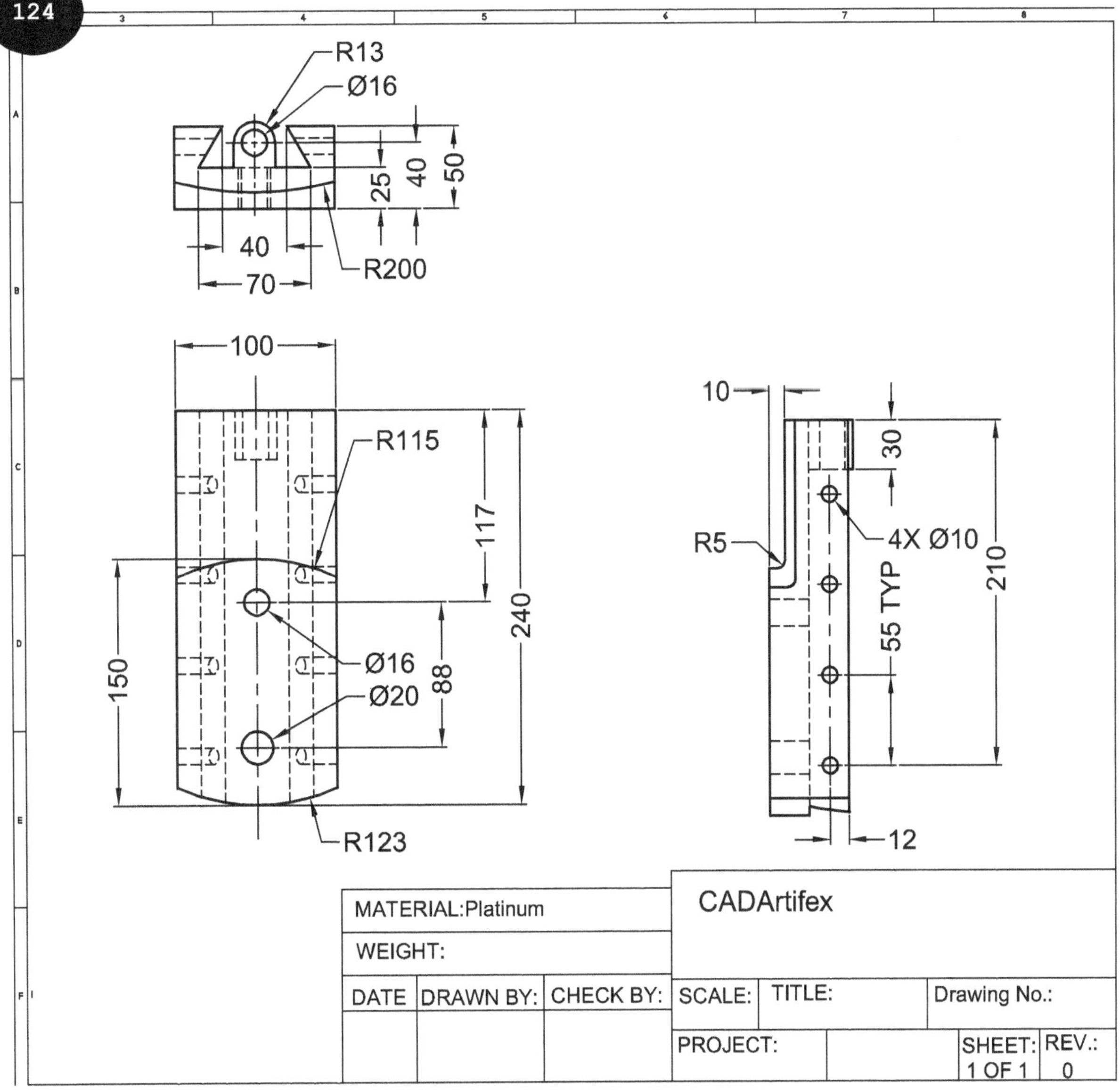

R13
Ø16
25
40
50
40
70
R200

100
R115
117
240
150
Ø16
88
Ø20
R123

10
30
R5
4X Ø10
55 TYP
210
12

MATERIAL:Platinum			CADArtifex		
WEIGHT:					
DATE	DRAWN BY:	CHECK BY:	SCALE:	TITLE:	Drawing No.:
			PROJECT:		SHEET: 1 OF 1	REV.: 0

Exercise 63.

Create the 3D model, as shown in Figure 125. Different views of the model and dimensions are shown in Figure 126. After creating the model, assign the Steel AISI 1020 107 HR material and calculate its mass properties. All dimensions are in mm.

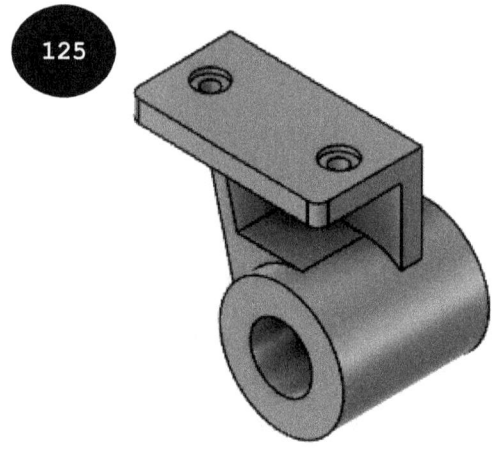

125

126

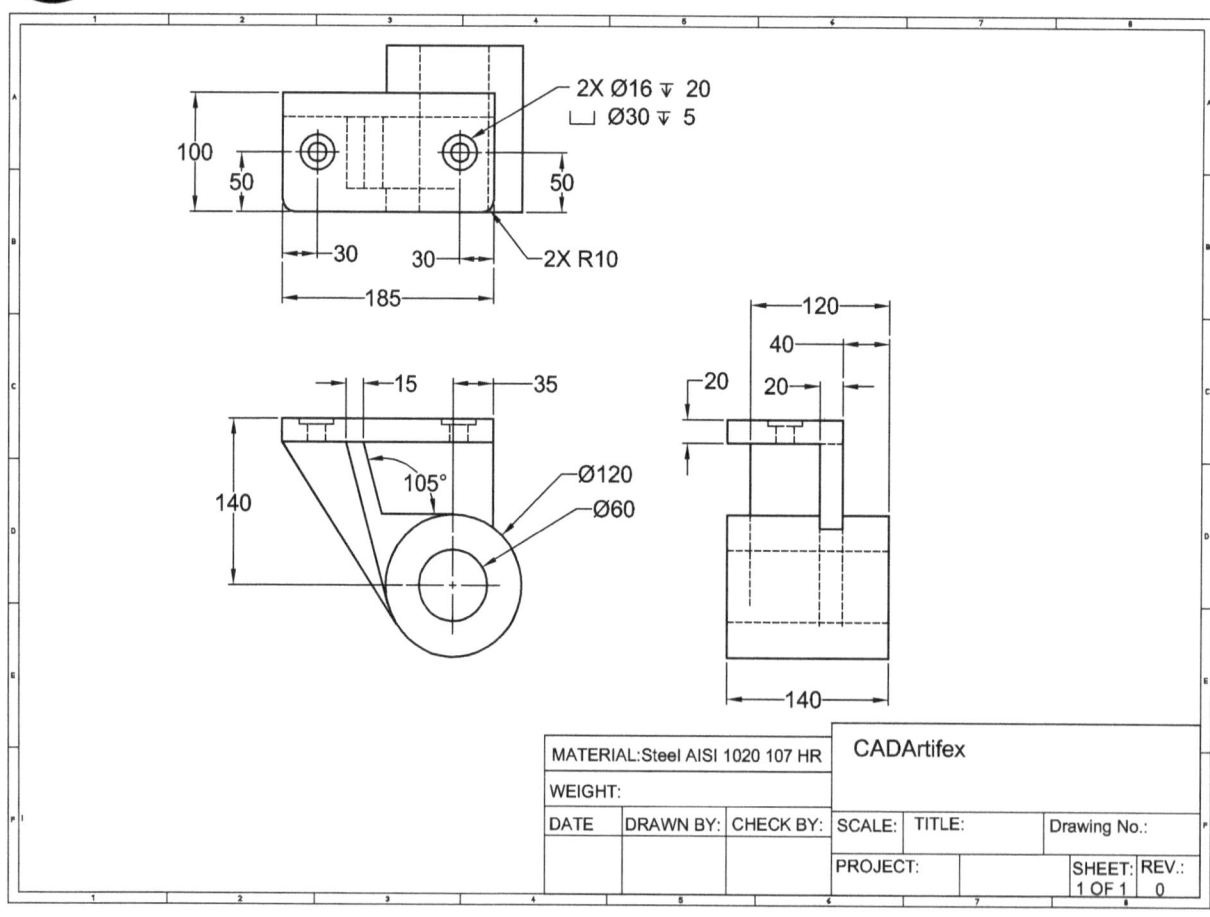

Exercise 64.

Create the 3D model, as shown in Figure 127. Different views of the model and dimensions are shown in Figure 128. After creating the model, assign the Stainless Steel material and calculate its mass properties. All dimensions are in mm.

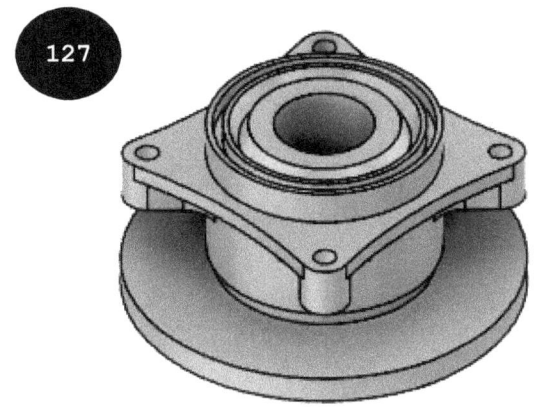

127

128

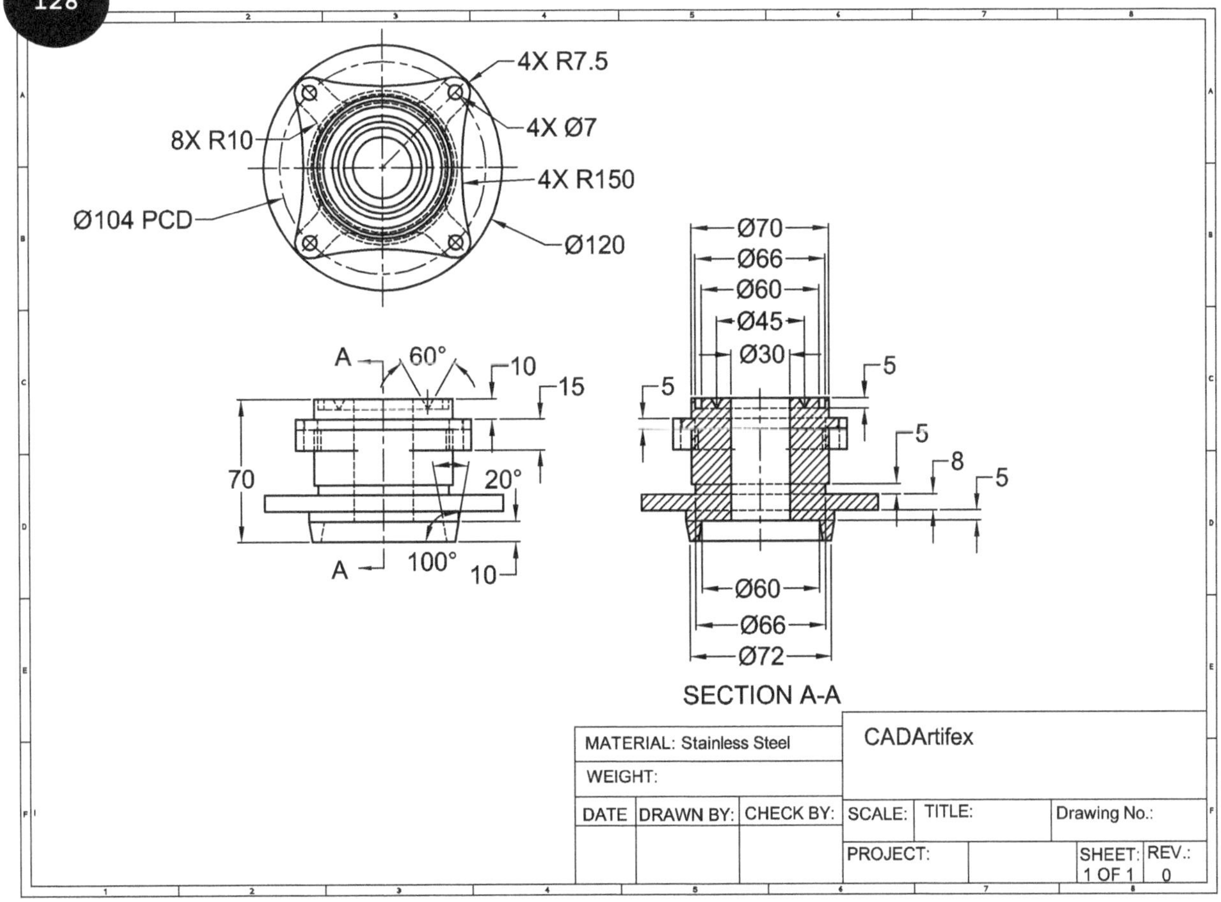

4X R7.5
4X Ø7
8X R10
4X R150
Ø104 PCD
Ø120

Ø70
Ø66
Ø60
Ø45
Ø30

A 60° 10
15
70 20°
A 100° 10

5
5
5
8
5

Ø60
Ø66
Ø72

SECTION A-A

MATERIAL: Stainless Steel			CADArtifex		
WEIGHT:					
DATE	DRAWN BY:	CHECK BY:	SCALE:	TITLE:	Drawing No.:
			PROJECT:		SHEET: REV.: 1 OF 1 0

Exercise 65.

Create the 3D model, as shown in Figure 129. Different views and dimensions are shown in Figure 130. After creating the model, assign the Platinum material and calculate its mass properties. All dimensions are in mm.

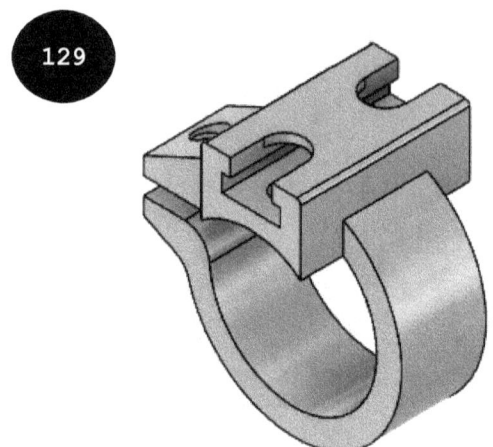

129

130

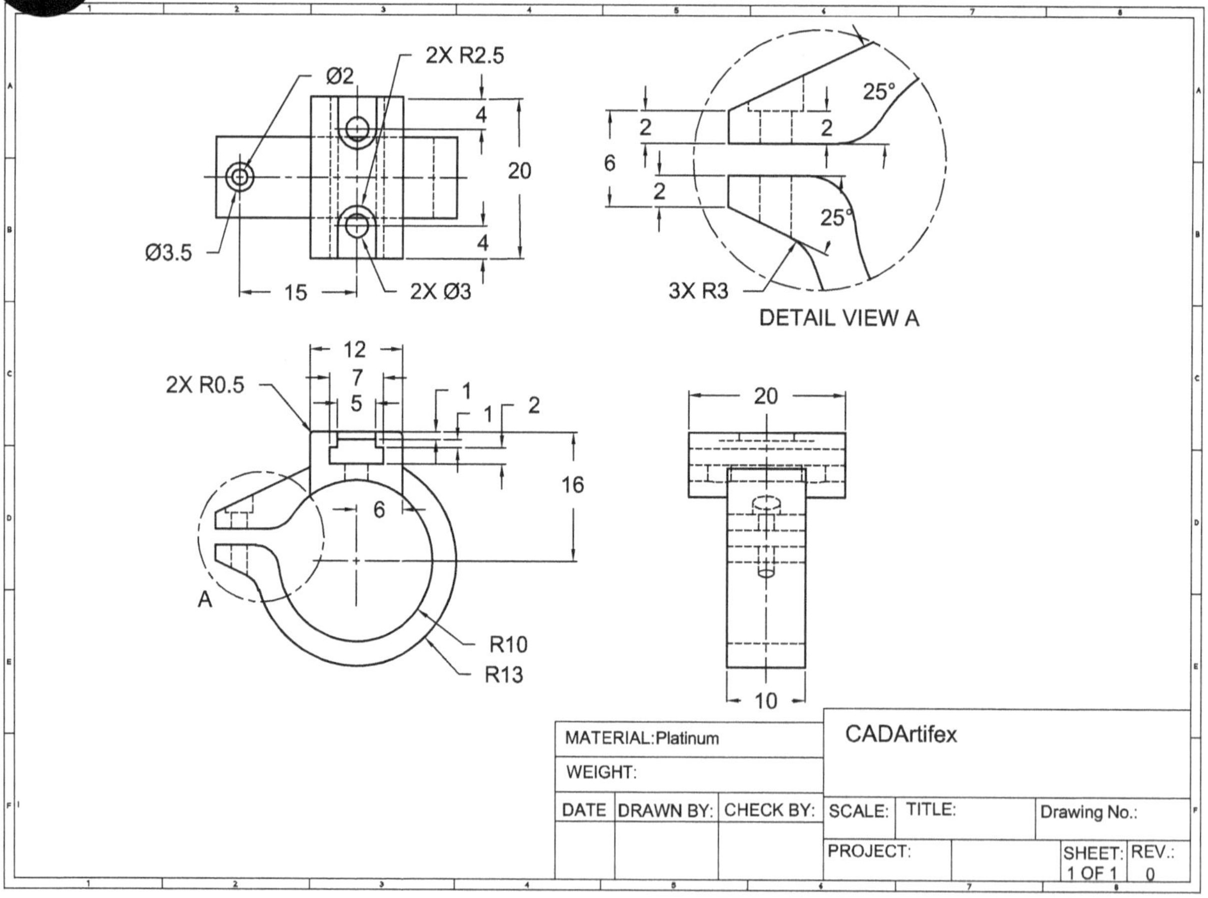

2X R2.5

Ø2

4

20

4

Ø3.5

15

2X Ø3

25°

2

2

6

2

25°

3X R3

DETAIL VIEW A

12

7

5

2X R0.5

1

1

2

16

6

A

R10

R13

20

10

MATERIAL:Platinum			CADArtifex			
WEIGHT:						
DATE	DRAWN BY:	CHECK BY:	SCALE:	TITLE:		Drawing No.:
			PROJECT:			SHEET: 1 OF 1 / REV.: 0

Exercise 66.

Create the 3D model, as shown in Figure 131. Different views of the model and dimensions are shown in Figure 132. After creating the model, assign the Stainless Steel material and calculate its mass properties. All dimensions are in mm.

131

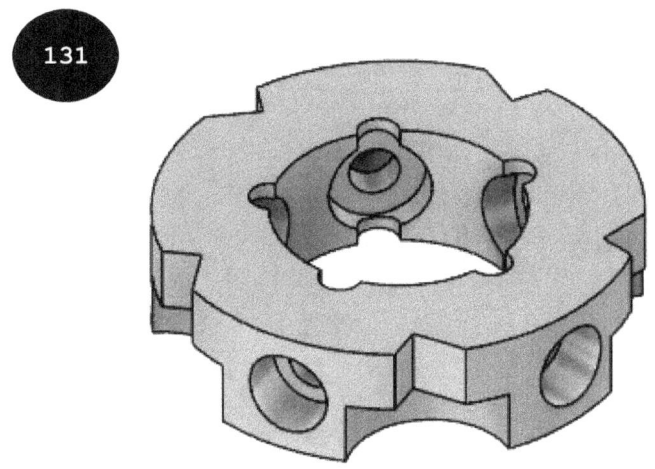

132

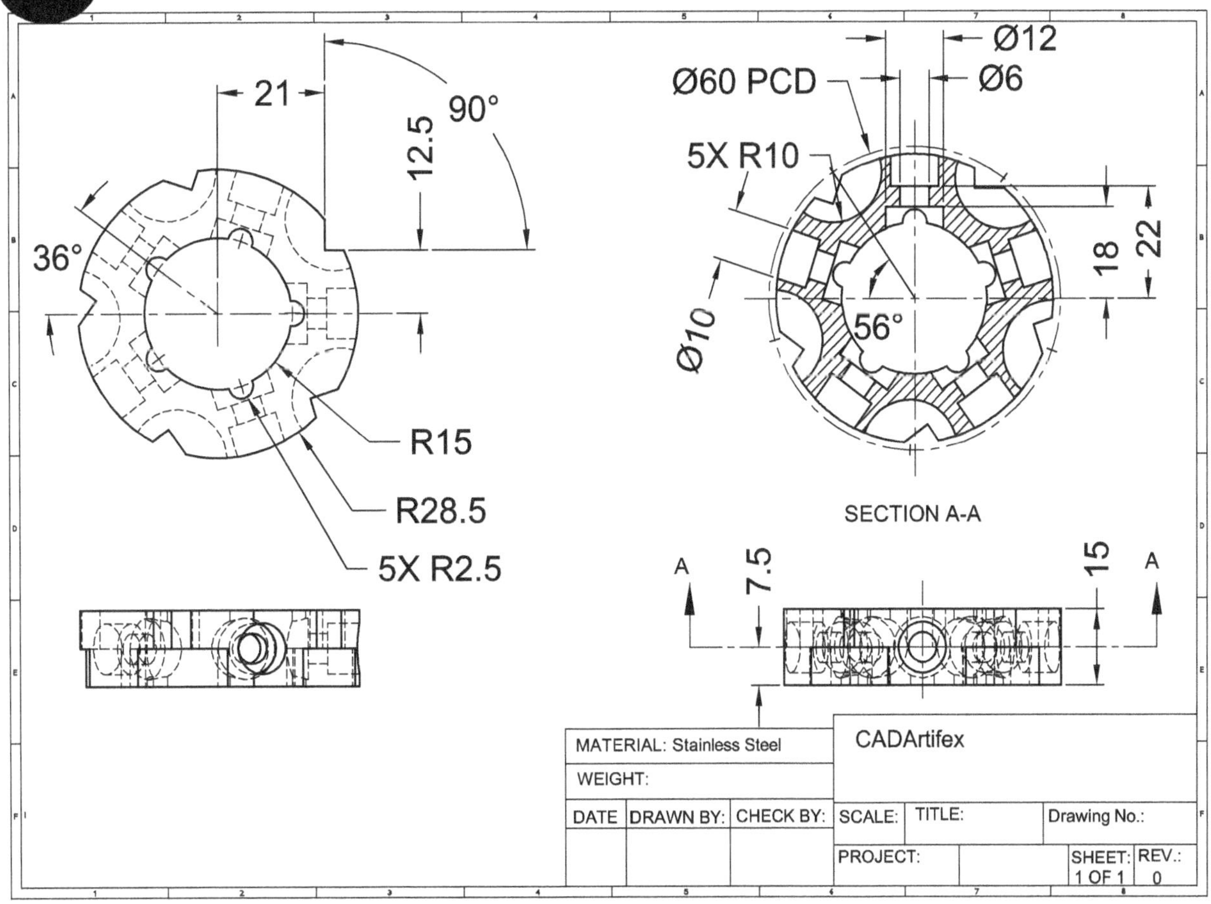

Ø12
Ø60 PCD
Ø6
5X R10
21
90°
12.5
36°
18
22
Ø10
56°
R15
R28.5
5X R2.5

SECTION A-A

A
7.5
15
A

MATERIAL: Stainless Steel			CADArtifex		
WEIGHT:					
DATE	DRAWN BY:	CHECK BY:	SCALE:	TITLE:	Drawing No.:
			PROJECT:		SHEET: 1 OF 1	REV.: 0

Exercise 67.

Create the 3D model, as shown in Figure 133. Different views of the model and dimensions are shown in Figure 134. After creating the model, assign the Stainless Steel AISI 304 material and calculate its mass properties. All dimensions are in mm.

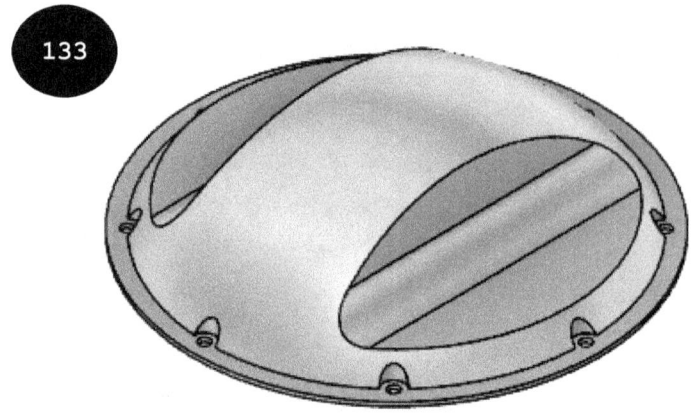

133

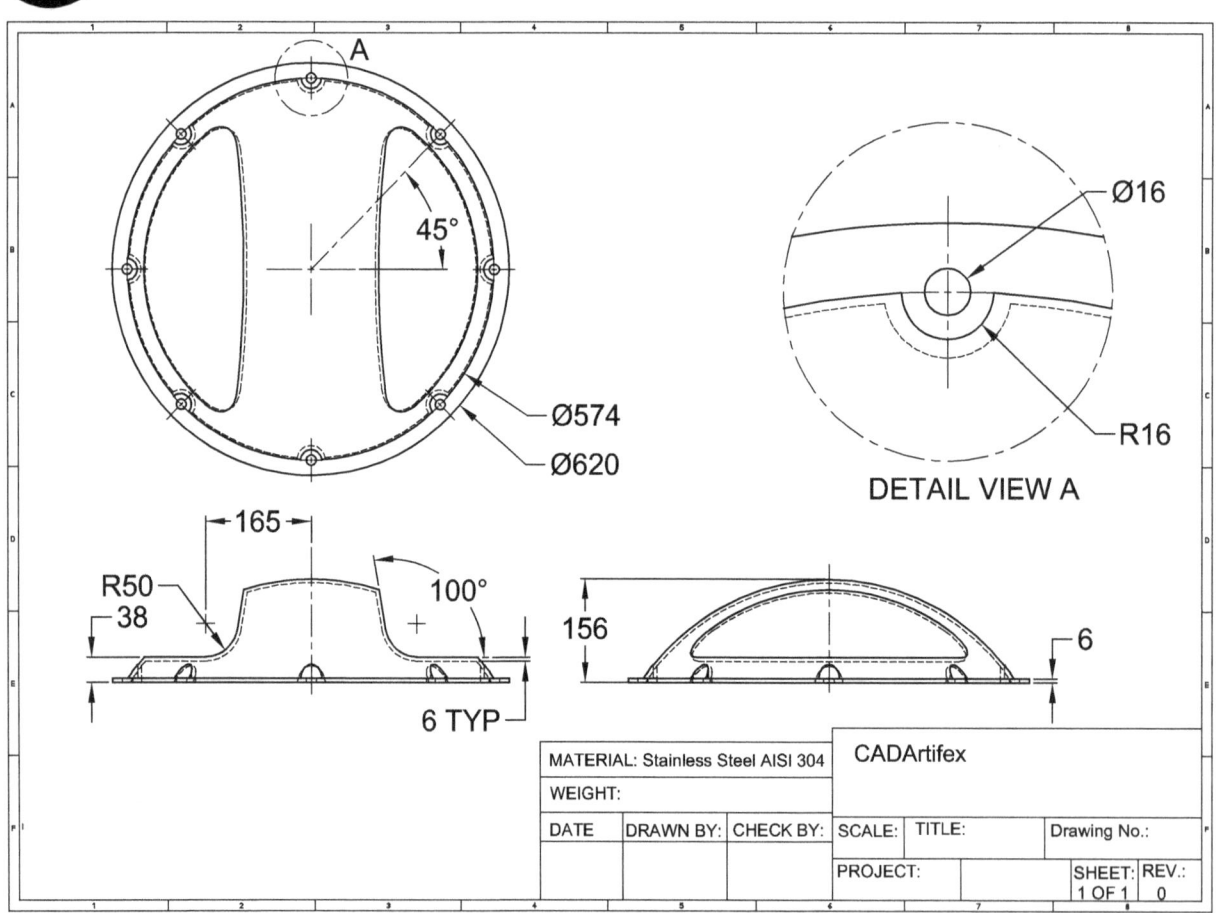

134

Exercise 68.

Create the 3D model, as shown in Figure 135. Different views of the model and dimensions are shown in Figure 136. After creating the model, assign the Steel AISI 1020 107 HR material and calculate its mass properties. All dimensions are in mm.

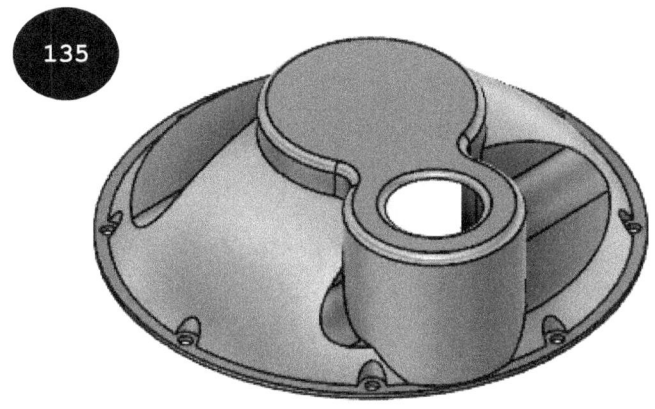

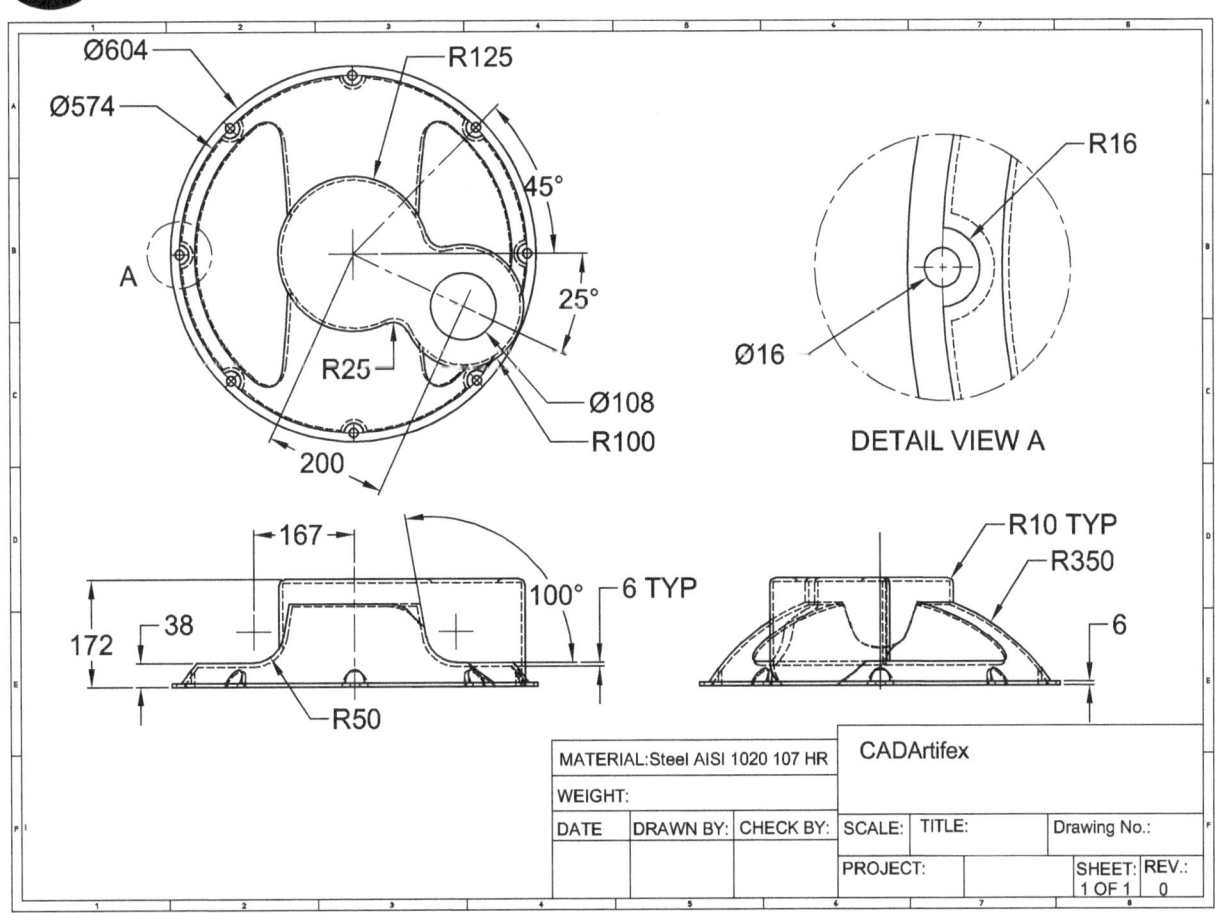

Exercise 69.

Create the 3D model, as shown in Figure 137. Different views of the model and dimensions are shown in Figure 138. After creating the model, assign the Stainless Steel AISI 304 material and calculate its mass properties. All dimensions are in mm.

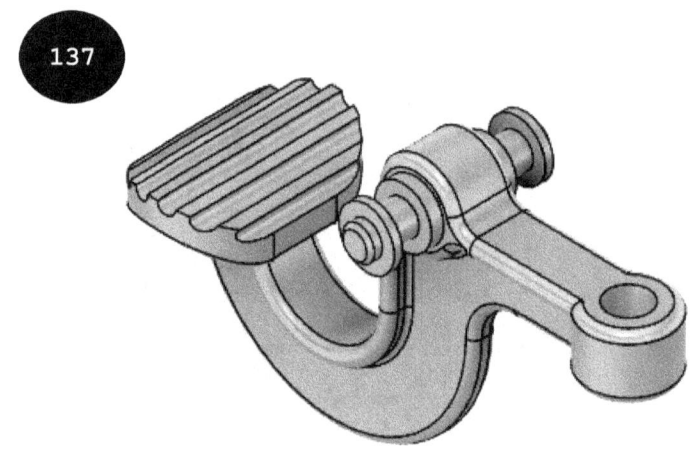

137

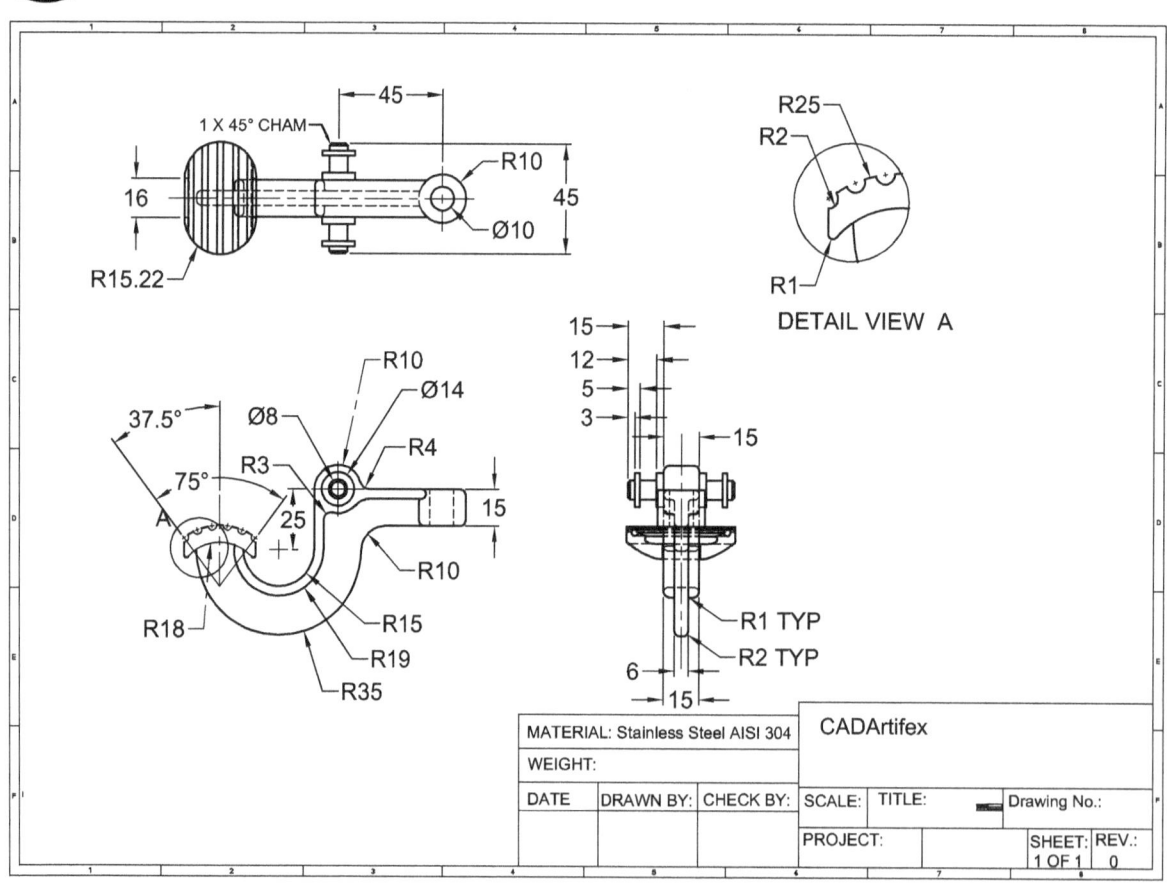

138

Exercise 70.

Create the 3D model, as shown in Figure 139. Different views of the model and dimensions are shown in Figure 140. After creating the model, assign the Platinum material and calculate its mass properties. All dimensions are in mm.

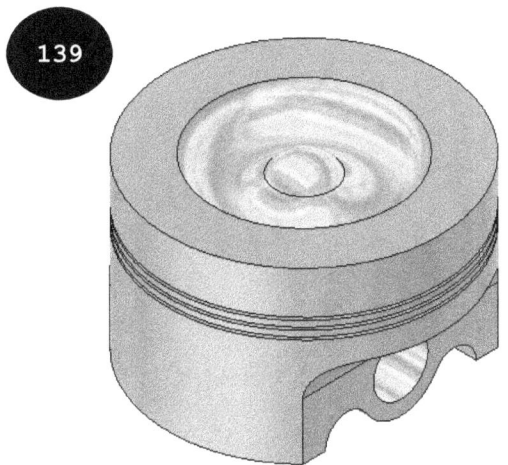

139

140

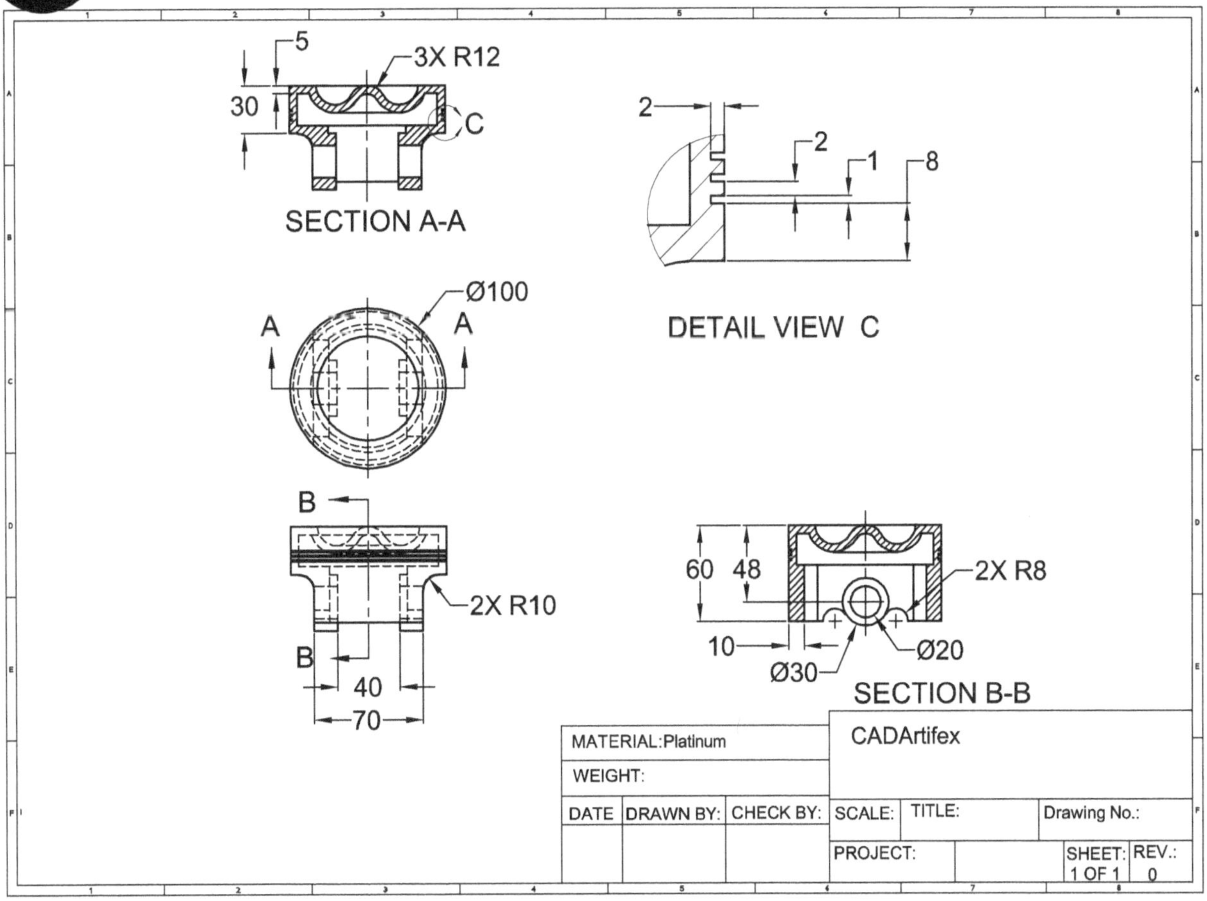

Exercise 71.

Create the 3D model, as shown in Figure 141. Different views of the model and dimensions are shown in Figure 142. After creating the model, assign the Stainless Steel AISI 304 material and calculate its mass properties. All dimensions are in mm.

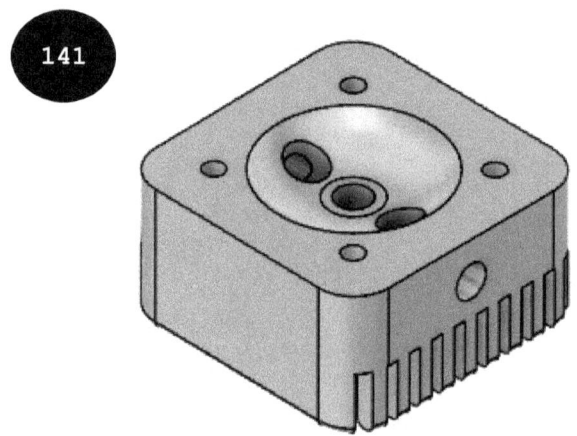

141

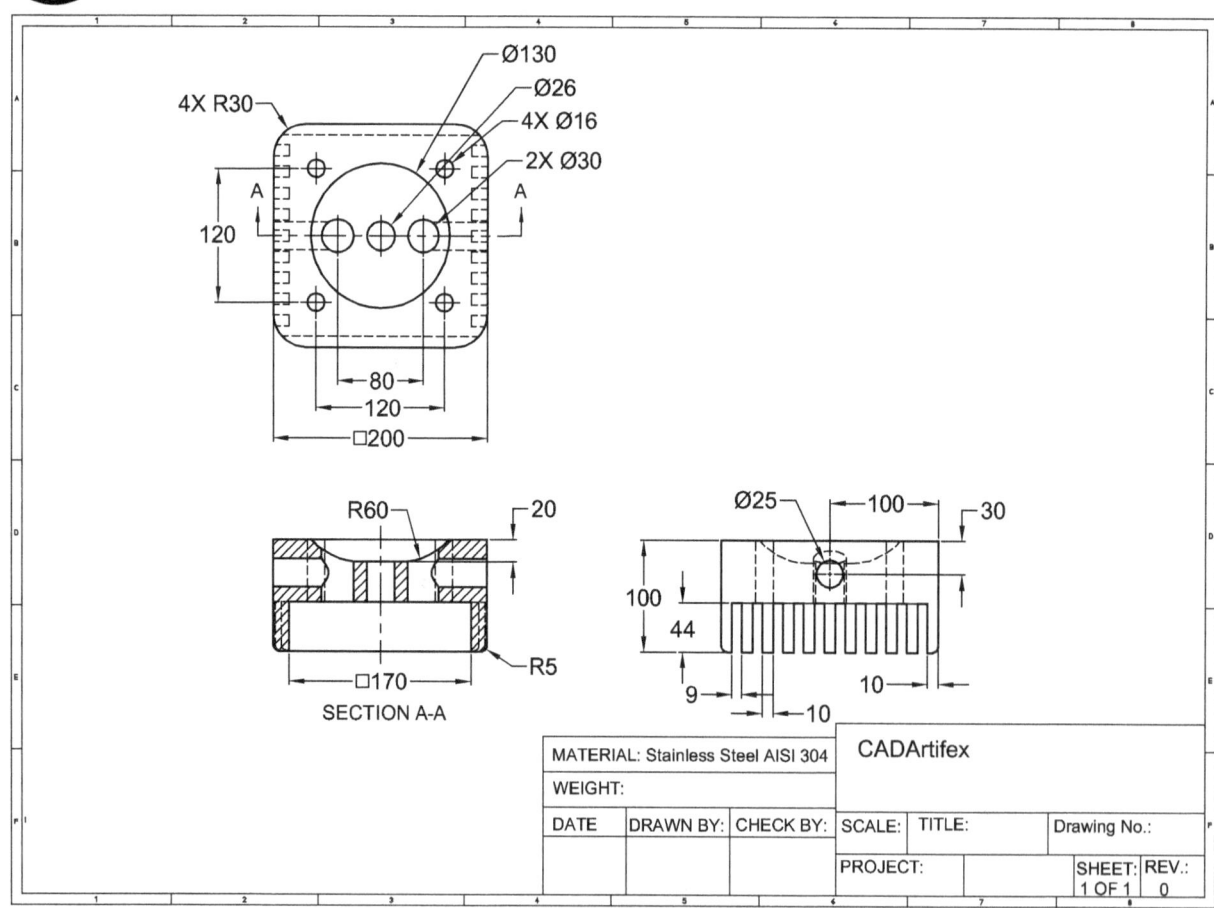

142

Exercise 72.

Create the 3D model, as shown in Figure 143. Different views of the model and dimensions are shown in Figure 144. After creating the model, assign the Stainless Steel AISI 304 material and calculate its mass properties. All dimensions are in mm.

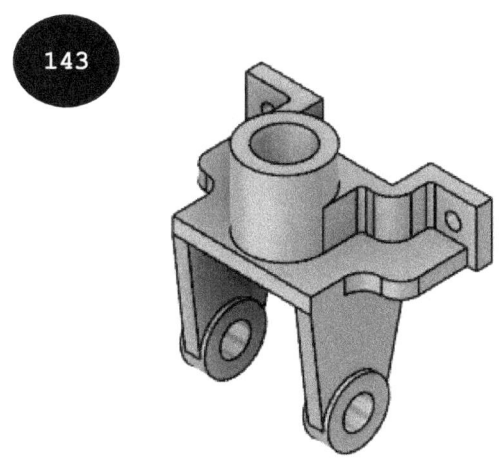

143

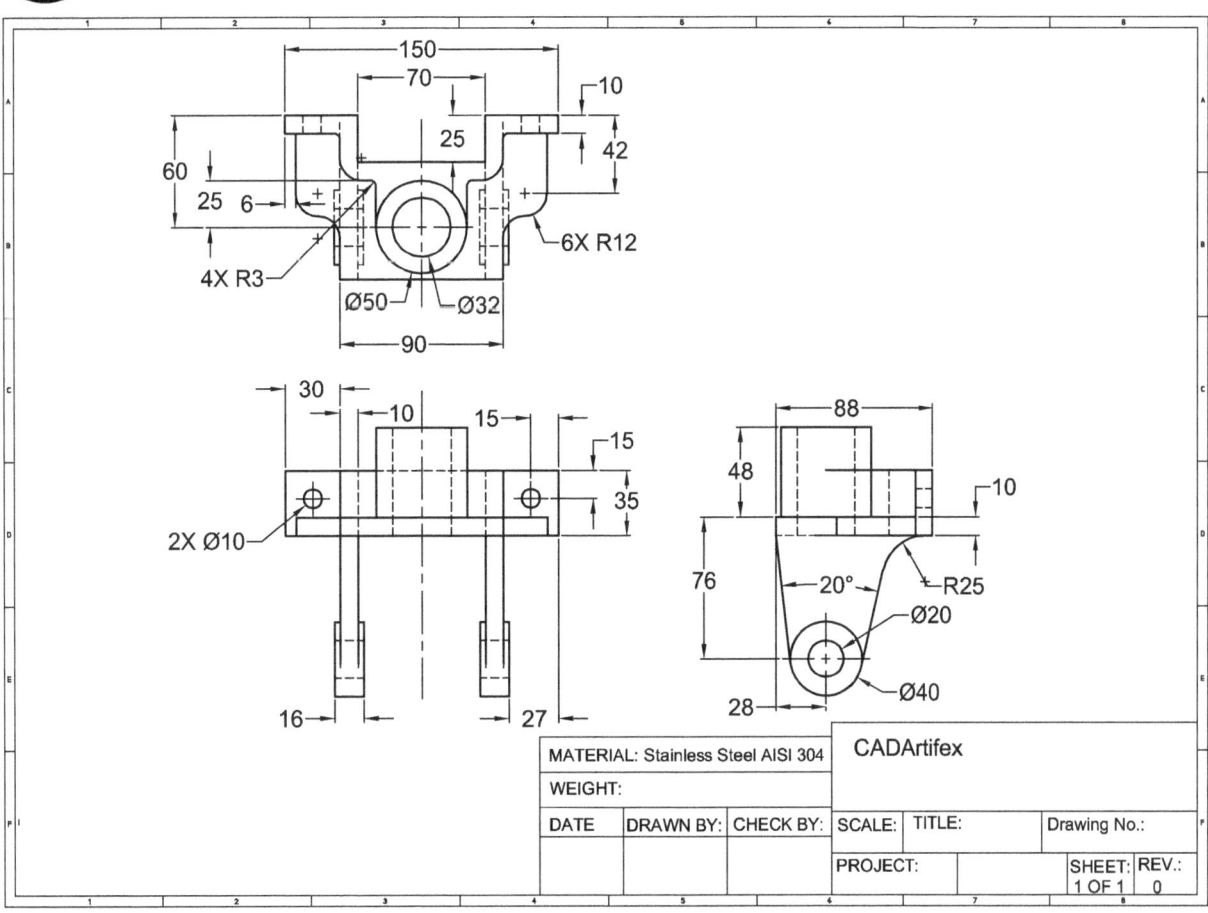

144

Exercise 73.

Create the 3D model, as shown in Figure 145. Different views of the model and dimensions are shown in Figure 146. After creating the model, assign the Steel AISI 1020 107 HR material and calculate its mass properties. All dimensions are in mm.

145

146

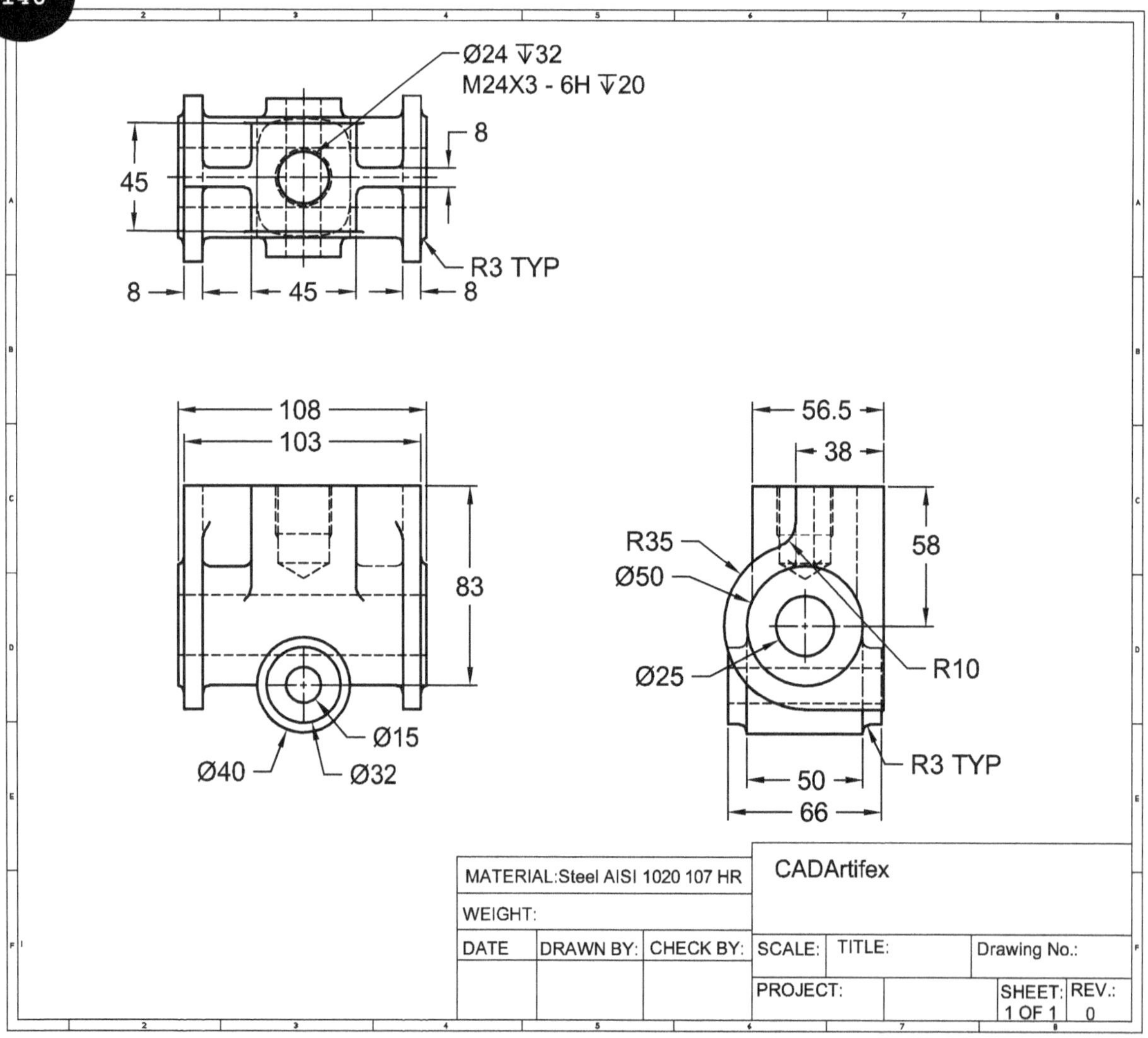

Ø24 ⊽32
M24X3 - 6H ⊽20

45

8

8 — 45 — 8

R3 TYP

108
103

83

Ø15
Ø40   Ø32

56.5
38

R35
Ø50

58

Ø25   R10

50
66

R3 TYP

MATERIAL:Steel AISI 1020 107 HR			CADArtifex			
WEIGHT:						
DATE	DRAWN BY:	CHECK BY:	SCALE:	TITLE:		Drawing No.:
			PROJECT:			SHEET: 1 OF 1 / REV.: 0

Exercise 74.

Create the 3D model, as shown in Figure 147. Different views of the model and dimensions are shown in Figure 148. After creating the model, assign the Platinum material and calculate its mass properties. All dimensions are in mm.

147

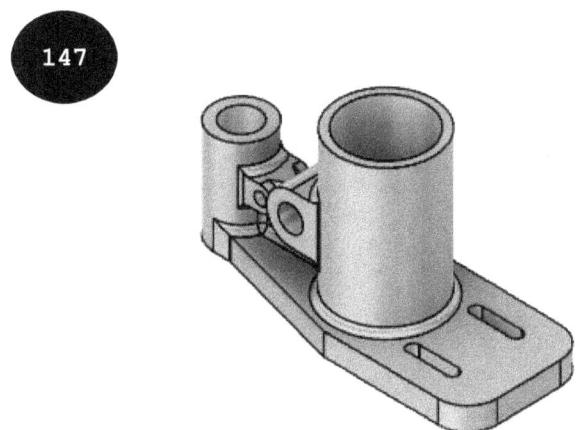

148

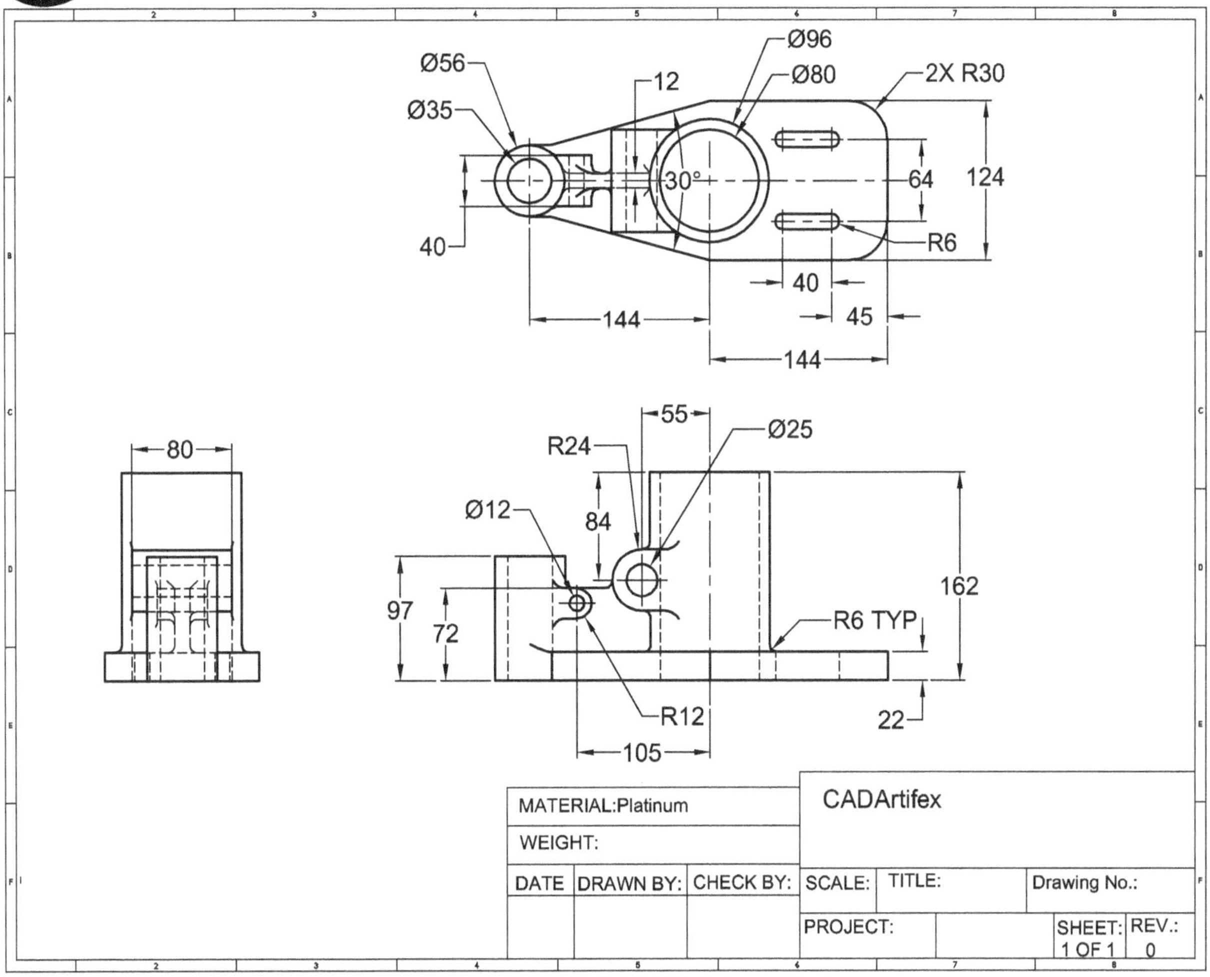

MATERIAL:Platinum	CADArtifex	
WEIGHT:		
DATE	DRAWN BY:	CHECK BY:	SCALE:	TITLE:	Drawing No.:
			PROJECT:		SHEET: REV.: 1 OF 1 0

Ø56
Ø35
Ø96
Ø80
2X R30
12
30°
40
124
64
144
R6
40
45
144

55
Ø25
R24
84
162
Ø12
80
97
72
R6 TYP
R12
22
105

Exercise 75.

Create the 3D model, as shown in Figure 149. Different views of the model and dimensions are shown in Figure 150. After creating the model, assign the Stainless Steel AISI 304 material and calculate its mass properties. All dimensions are in mm.

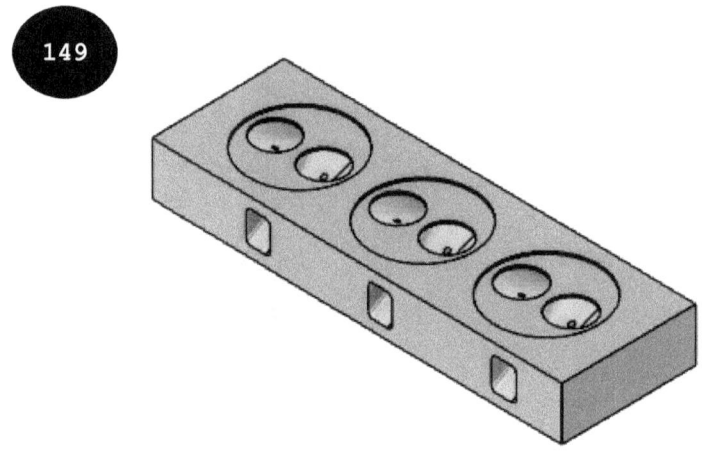

149

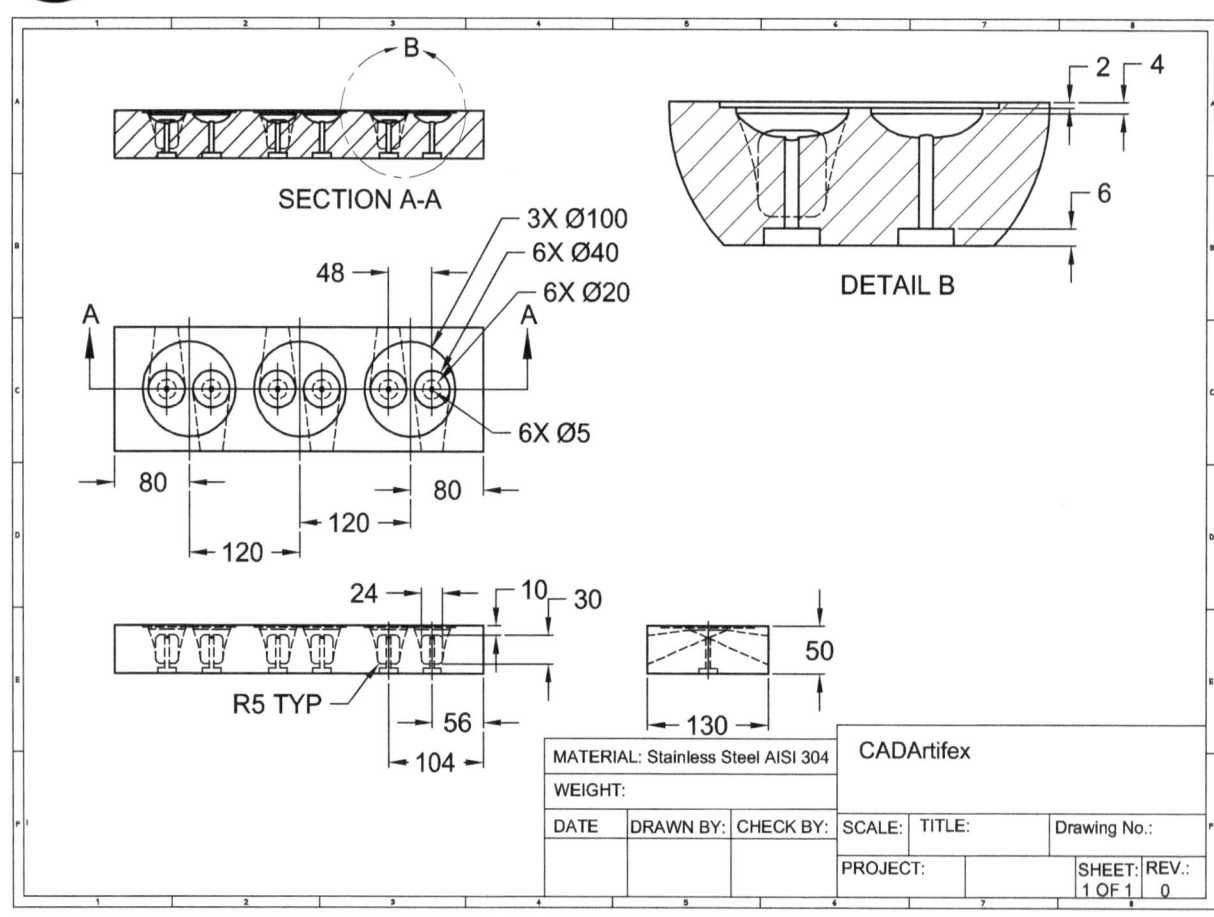

150

Exercise 76.

Create the 3D model, as shown in Figure 151. Different views of the model and dimensions are shown in Figure 152. After creating the model, assign the Platinum material and calculate its mass properties. All dimensions are in mm.

151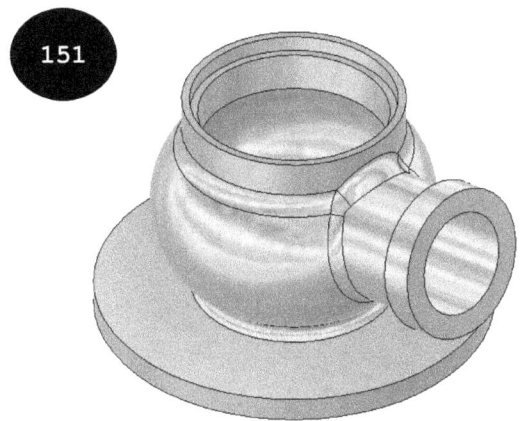

152

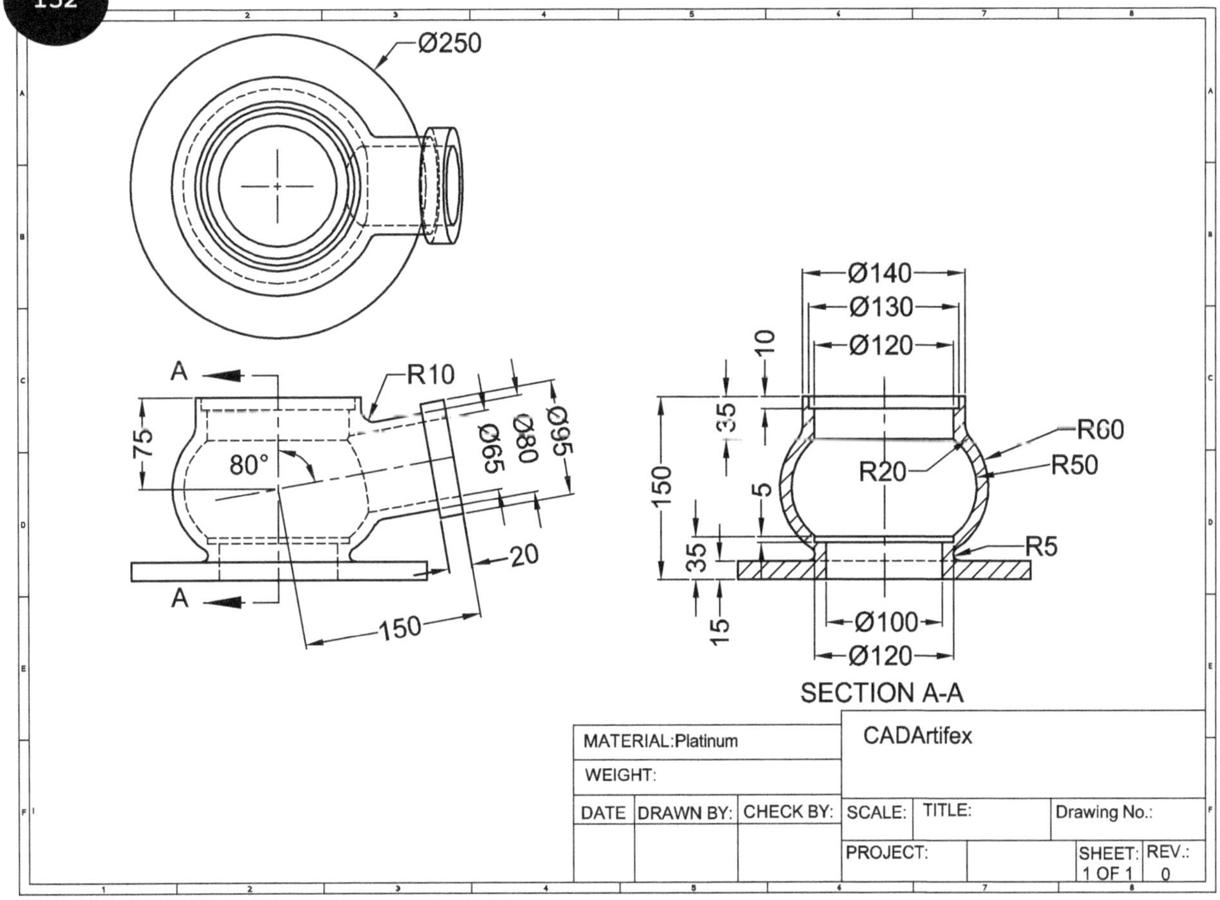

Ø250

Ø140
Ø130
Ø120

10

35

R10

Ø65
Ø80
Ø95

75

80°

150

R60
R50
R20

5

R5

35

20

150

15

Ø100
Ø120

SECTION A-A

MATERIAL:Platinum	CADArtifex			
WEIGHT:				
DATE	DRAWN BY:	CHECK BY:	SCALE:	TITLE:	Drawing No.:
			PROJECT:		SHEET: 1 OF 1	REV.: 0

Exercise 77.

Create the 3D model, as shown in Figure 153. Different views of the model and dimensions are shown in Figure 154. After creating the model, assign the Stainless Steel AISI 304 material and calculate its mass properties. All dimensions are in mm.

153

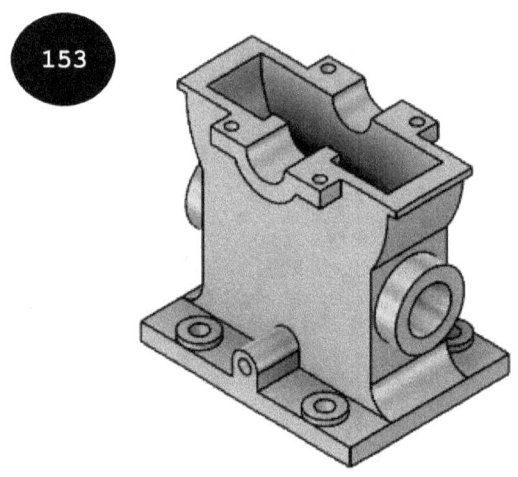

154

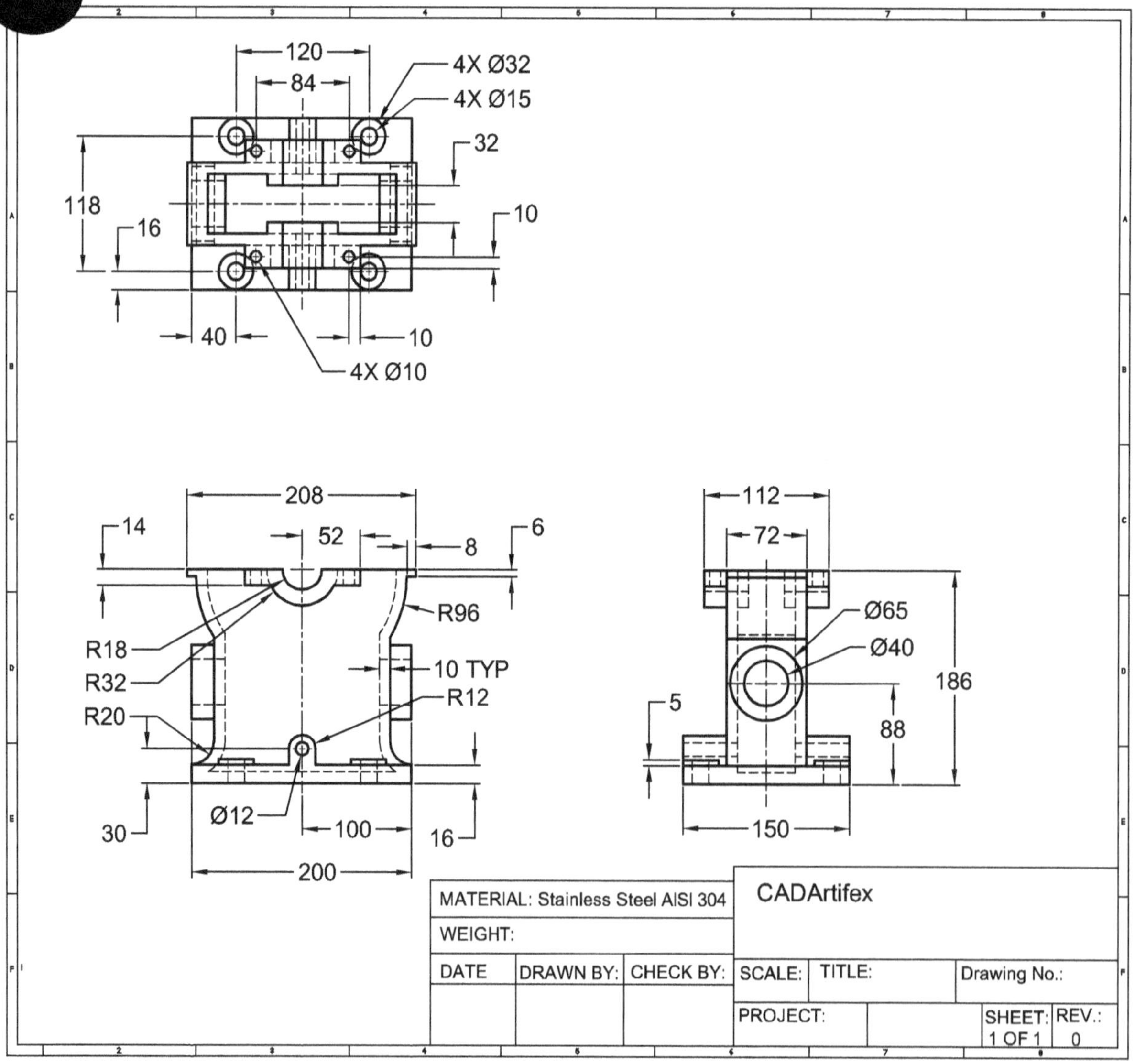

120
84
4X Ø32
4X Ø15
32
118
16
10
40
10
4X Ø10

14
208
52
8
6
R96
R18
R32
10 TYP
R20
R12
Ø12
100
16
30
200

112
72
Ø65
Ø40
5
186
88
150

MATERIAL: Stainless Steel AISI 304			CADArtifex			
WEIGHT:						
DATE	DRAWN BY:	CHECK BY:	SCALE:	TITLE:		Drawing No.:
			PROJECT:			SHEET: 1 OF 1 / REV.: 0

Exercise 78.

Create the 3D model, as shown in Figure 155. Different views of the model and dimensions are shown in Figure 156. After creating the model, assign the Steel AISI 1020 107 HR material and calculate its mass properties. All dimensions are in mm.

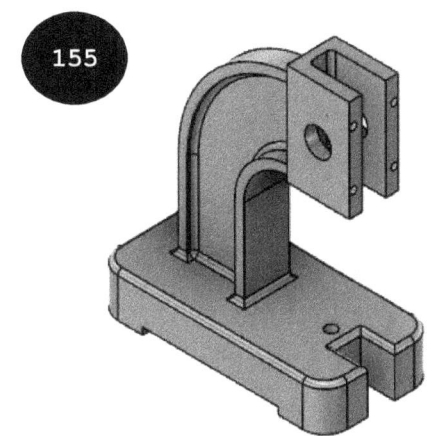

155

156

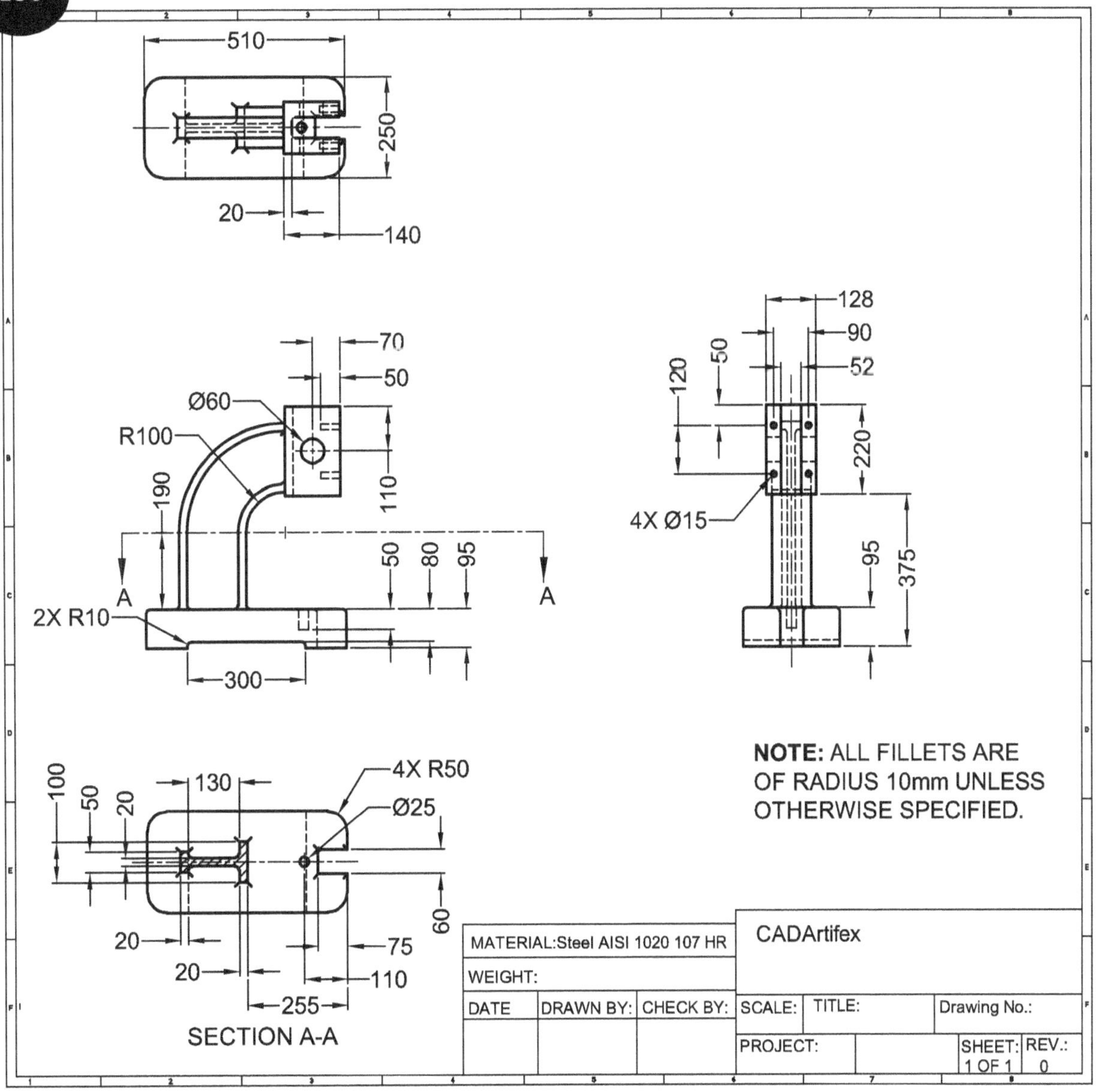

510
250
20
140

70
50
Ø60
R100
190
110
50
80
95
A
A
2X R10
300

128
90
52
120
50
220
4X Ø15
95
375

100
50
20
130
4X R50
Ø25
20
75
20
110
255
60

SECTION A-A

NOTE: ALL FILLETS ARE OF RADIUS 10mm UNLESS OTHERWISE SPECIFIED.

MATERIAL:Steel AISI 1020 107 HR				CADArtifex			
WEIGHT:							
DATE	DRAWN BY:	CHECK BY:	SCALE:	TITLE:		Drawing No.:	
			PROJECT:			SHEET: 1 OF 1	REV.: 0

Exercise 79.

Create the 3D model, as shown in Figure 157. Different views of the model and dimensions are shown in Figure 158. After creating the model, assign the Stainless Steel material and calculate its mass properties. All dimensions are in mm.

157

158

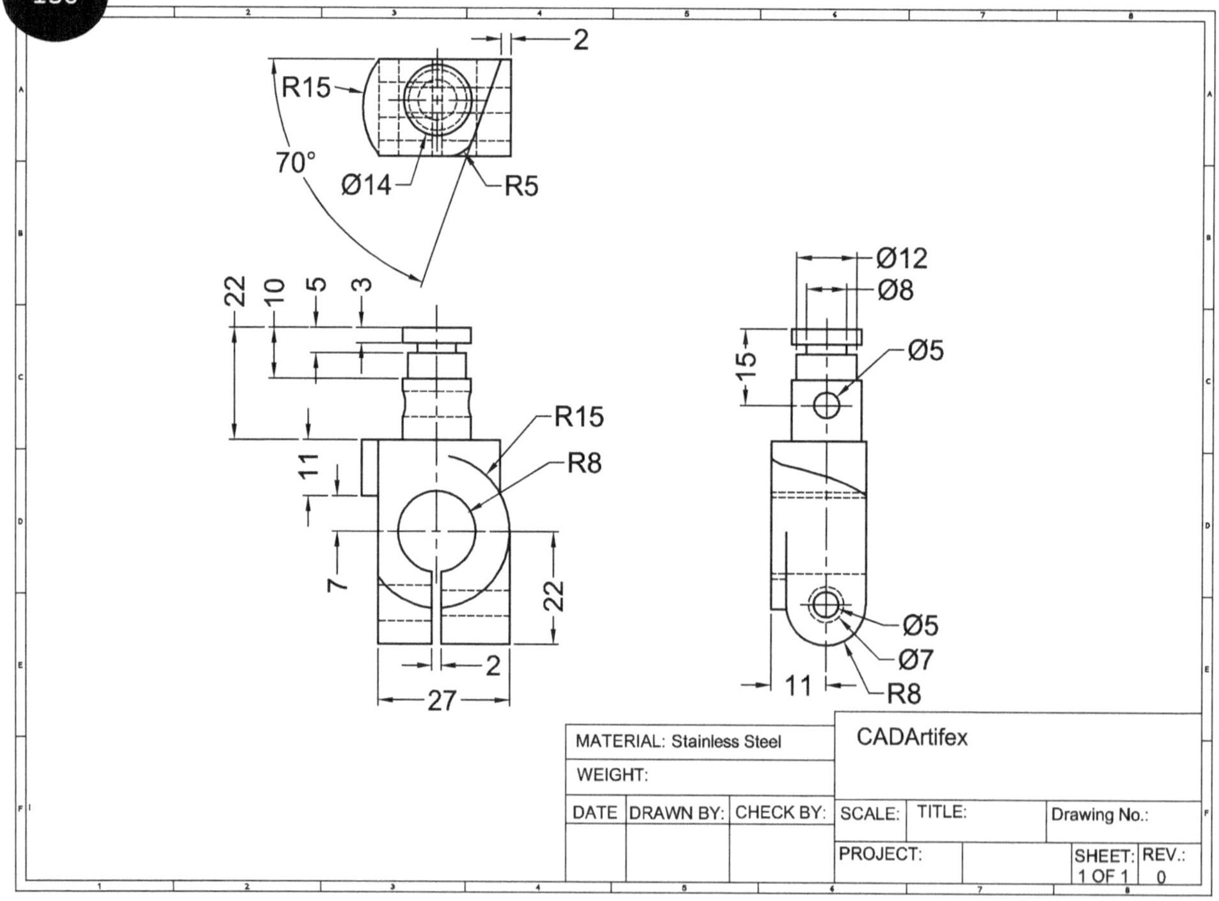

R15

70°

Ø14   R5

2

22 10 5 3

R15
R8

11

7

2

27

Ø12
Ø8

Ø5

15

Ø5
Ø7
11   R8

22

MATERIAL: Stainless Steel			CADArtifex			
WEIGHT:						
DATE	DRAWN BY:	CHECK BY:	SCALE:	TITLE:		Drawing No.:
			PROJECT:			SHEET: 1 OF 1 / REV.: 0

Exercise 80.

Create the 3D model, as shown in Figure 159. Different views of the model and dimensions are shown in Figure 160. After creating the model, assign the Platinum material and calculate its mass properties. All dimensions are in mm.

159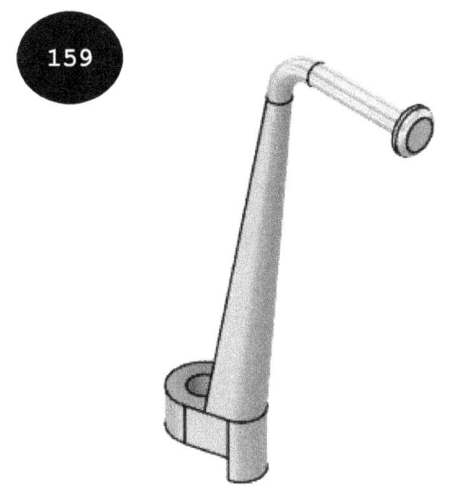

160

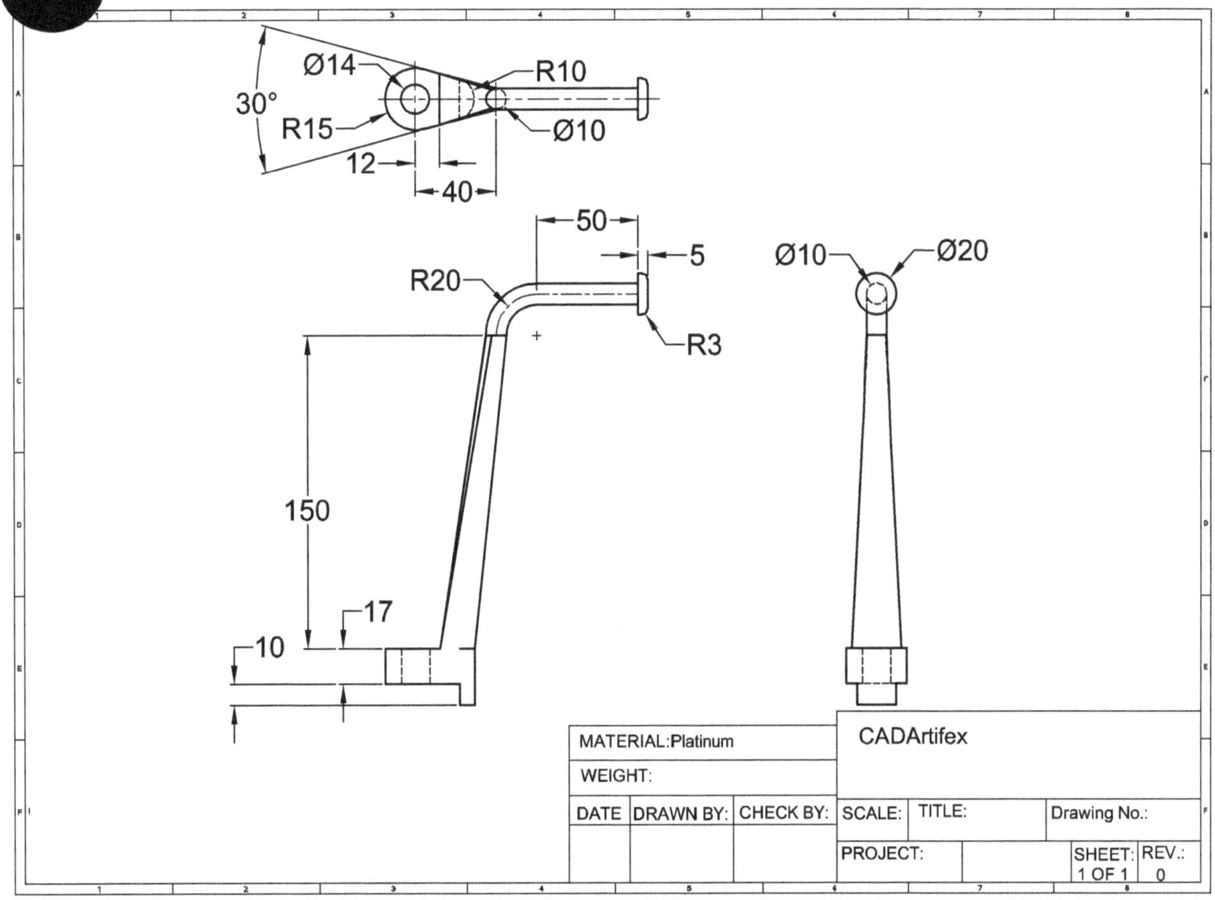

Ø14 R10
30° R15 Ø10
12
40

50
5
R20
R3
Ø10 Ø20

150

17
10

MATERIAL:Platinum			CADArtifex			
WEIGHT:						
DATE	DRAWN BY:	CHECK BY:	SCALE:	TITLE:		Drawing No.:
			PROJECT:			SHEET: 1 OF 1	REV.: 0

Exercise 81.

Create the 3D model, as shown in Figure 161. Different views of the model and dimensions are shown in Figure 162. After creating the model, assign the Stainless Steel AISI 304 material and calculate its mass properties. All dimensions are in mm.

161

162

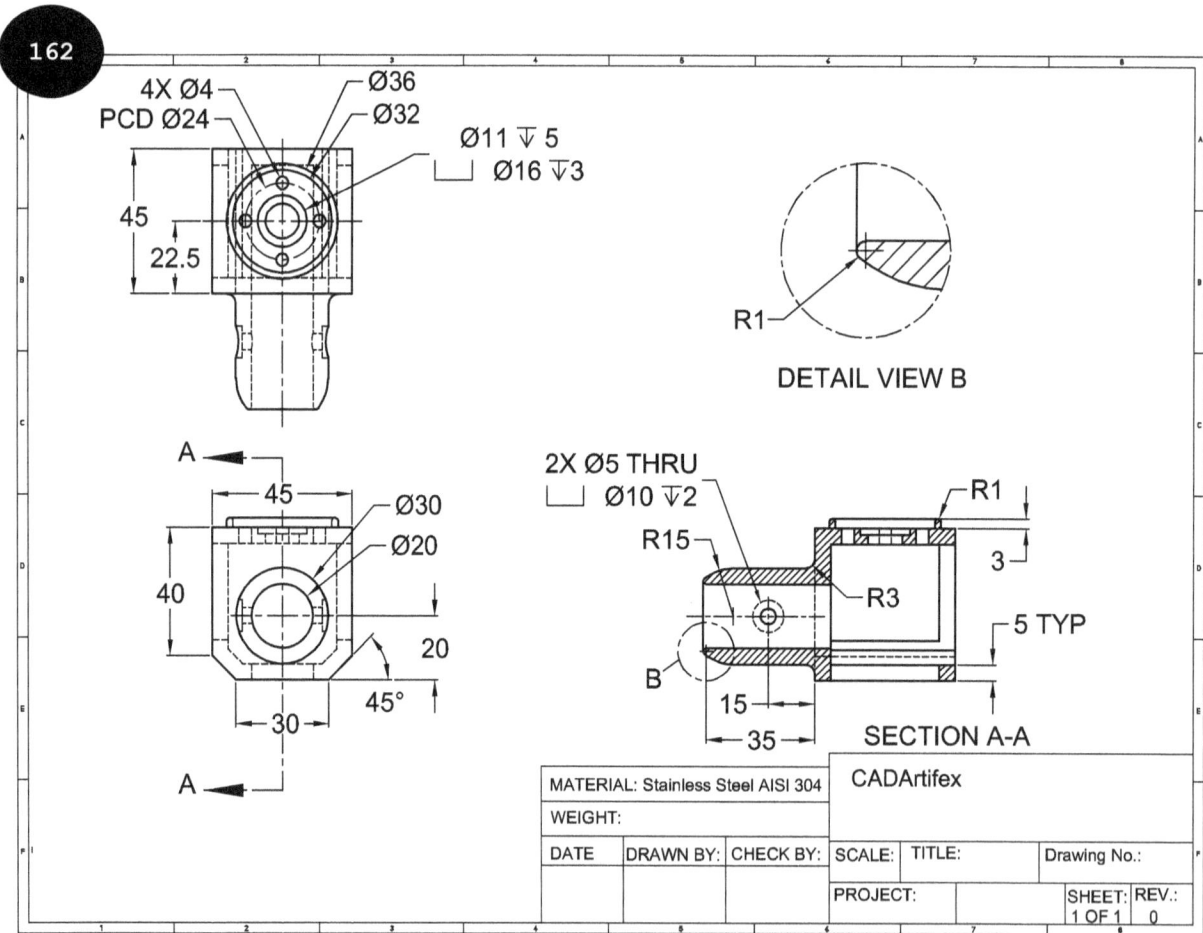

Exercise 82.

Create the 3D model, as shown in Figure 163. Different views of the model and dimensions are shown in Figure 164. After creating the model, assign the Stainless Steel material and calculate its mass properties. All dimensions are in mm.

163

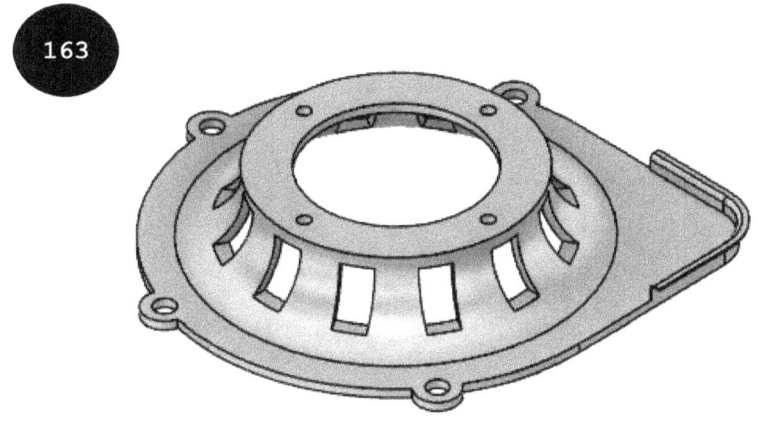

164

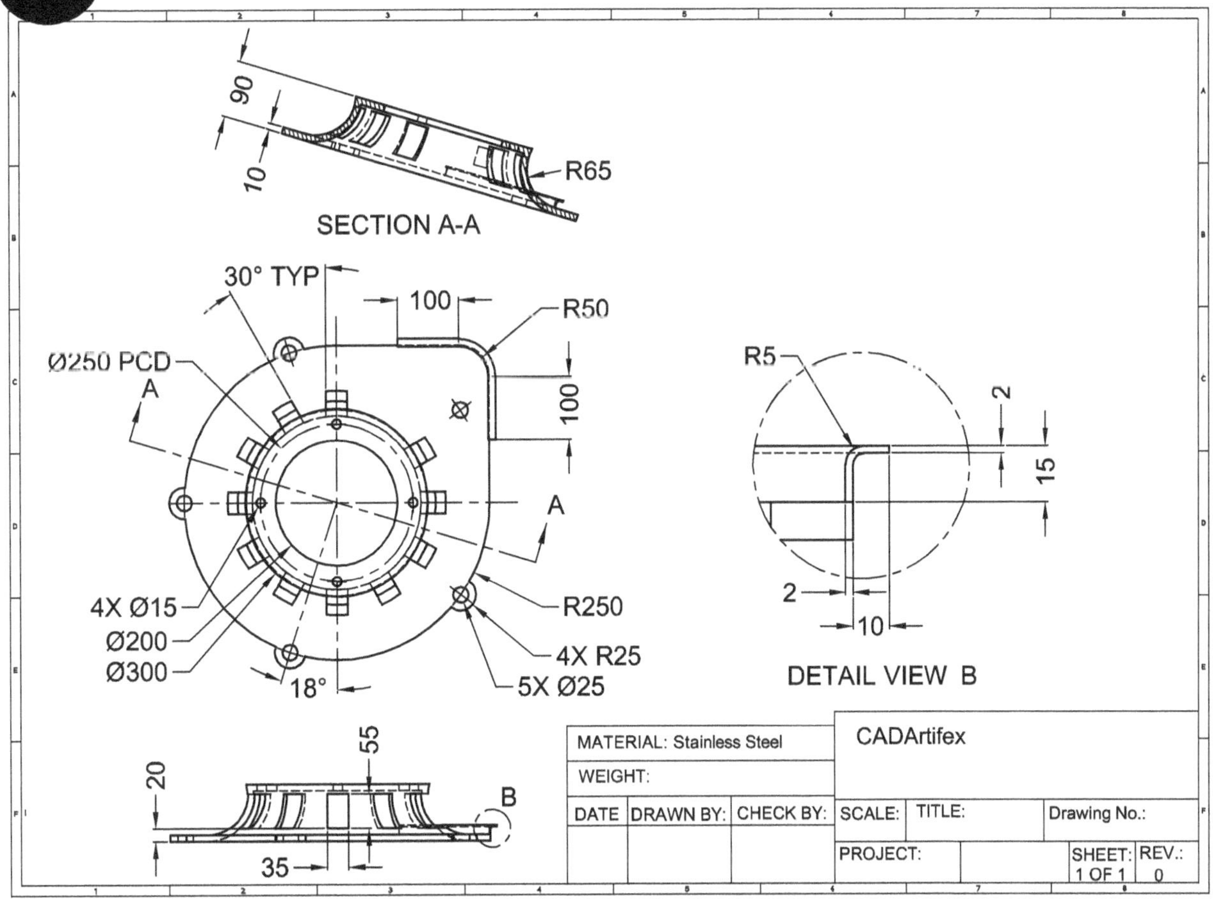

SECTION A-A

90
10
R65

30° TYP
Ø250 PCD
A
100
R50
100
A
R250
4X Ø15
Ø200
Ø300
18°
4X R25
5X Ø25

R5
2
15
2
10

DETAIL VIEW B

55
20
35
B

MATERIAL: Stainless Steel		CADArtifex			
WEIGHT:					
DATE	DRAWN BY:	CHECK BY:	SCALE:	TITLE:	Drawing No.:
			PROJECT:		SHEET: 1 OF 1	REV.: 0

Exercise 83.

Create the 3D model, as shown in Figure 165. Different views of the model and dimensions are shown in Figure 166. After creating the model, assign the Steel AISI 1020 107 HR material and calculate its mass properties. All dimensions are in mm.

165

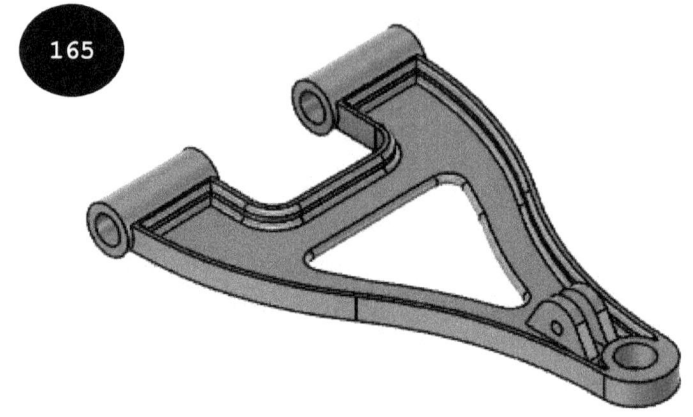

166

MATERIAL:Steel AISI 1020 107 HR		CADArtifex		
WEIGHT:				
DATE	DRAWN BY:	CHECK BY:	SCALE:	TITLE:	Drawing No.:
			PROJECT:		SHEET: 1 OF 1	REV.: 0

Exercise 84.

Create the 3D model, as shown in Figure 167. Different views of the model and dimensions are shown in Figure 168. After creating the model, assign the Stainless Steel material and calculate its mass properties. All dimensions are in mm.

167

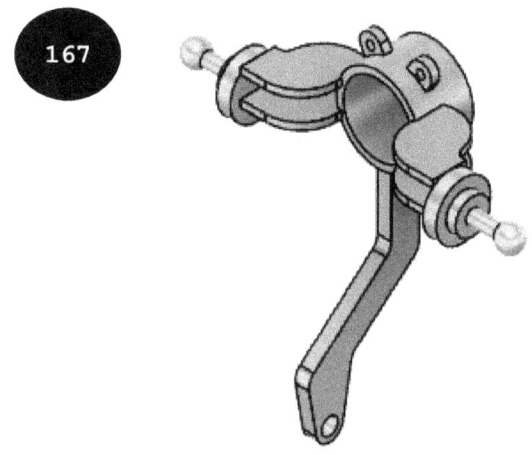

168

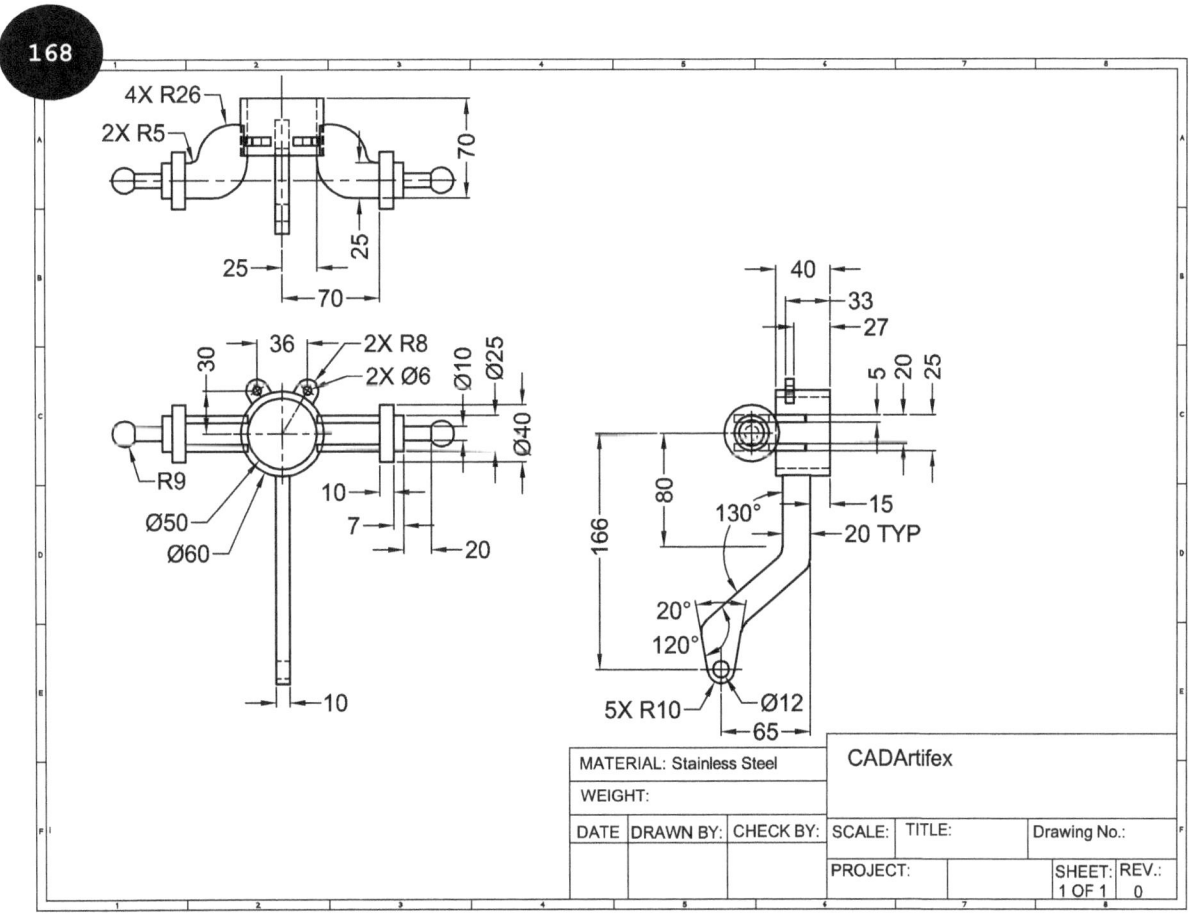

MATERIAL: Stainless Steel			CADArtifex			
WEIGHT:						
DATE	DRAWN BY:	CHECK BY:	SCALE:	TITLE:		Drawing No.:
			PROJECT:			SHEET: REV.:
						1 OF 1 0

Exercise 85.

Create the 3D model, as shown in Figure 169. Different views of the model and dimensions are shown in Figure 170. After creating the model, assign the Stainless Steel AISI 304 material and calculate its mass properties. All dimensions are in mm.

169

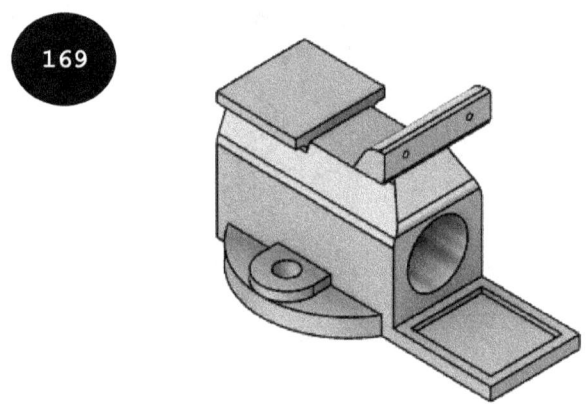

170

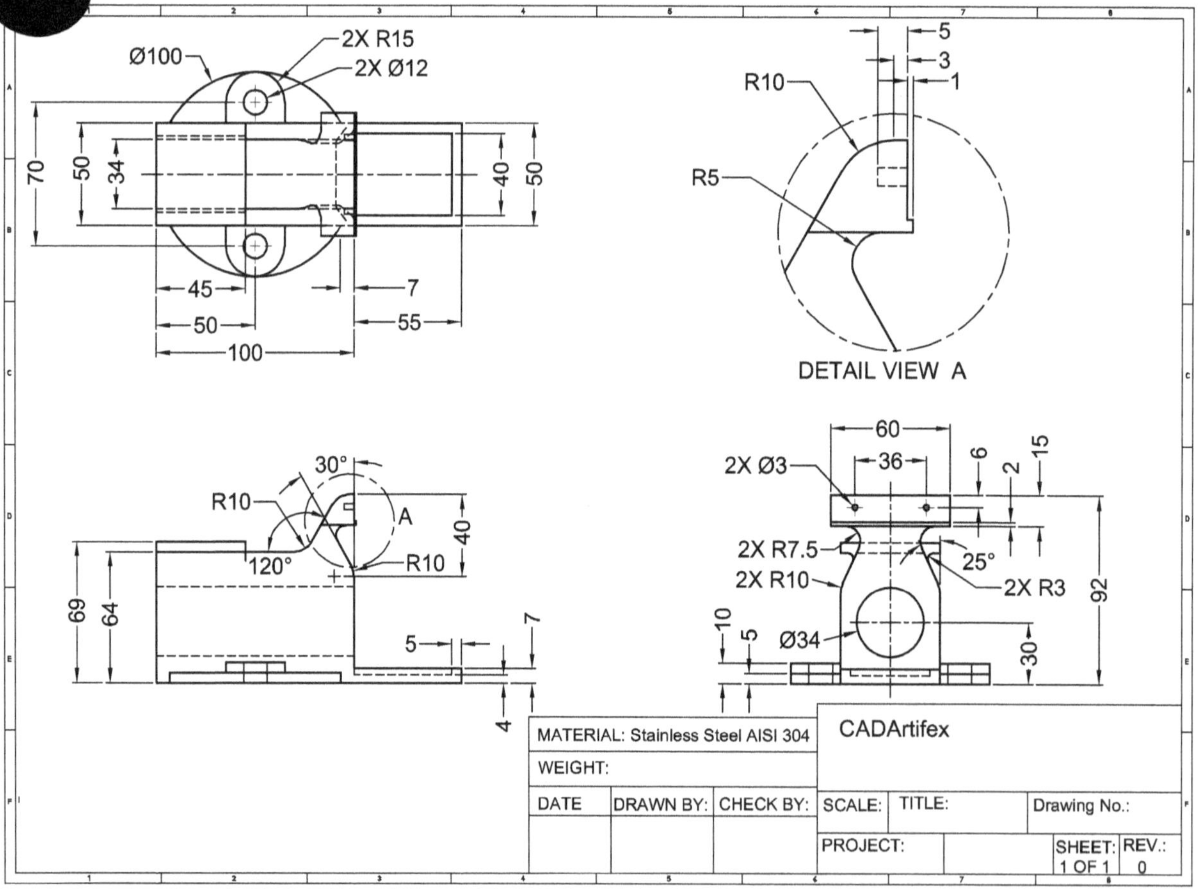

2X R15
Ø100
2X Ø12
70
50
34
40
50
45
7
50
55
100

R10
5
3
1
R5
DETAIL VIEW A

30°
R10
A
120°
R10
40
69
64
5
7
4

60
2X Ø3
36
6
2
15
2X R7.5
25°
2X R10
2X R3
10
5
Ø34
92
30

MATERIAL: Stainless Steel AISI 304		CADArtifex		
WEIGHT:				
DATE	DRAWN BY:	CHECK BY:	SCALE:	TITLE:	Drawing No.:
			PROJECT:		SHEET: REV.: 1 OF 1 0

Exercise 86.

Create the 3D model, as shown in Figure 171. After creating the model, assign the Steel AISI 1020 107 HR material and calculate its mass properties. All dimensions are in mm.

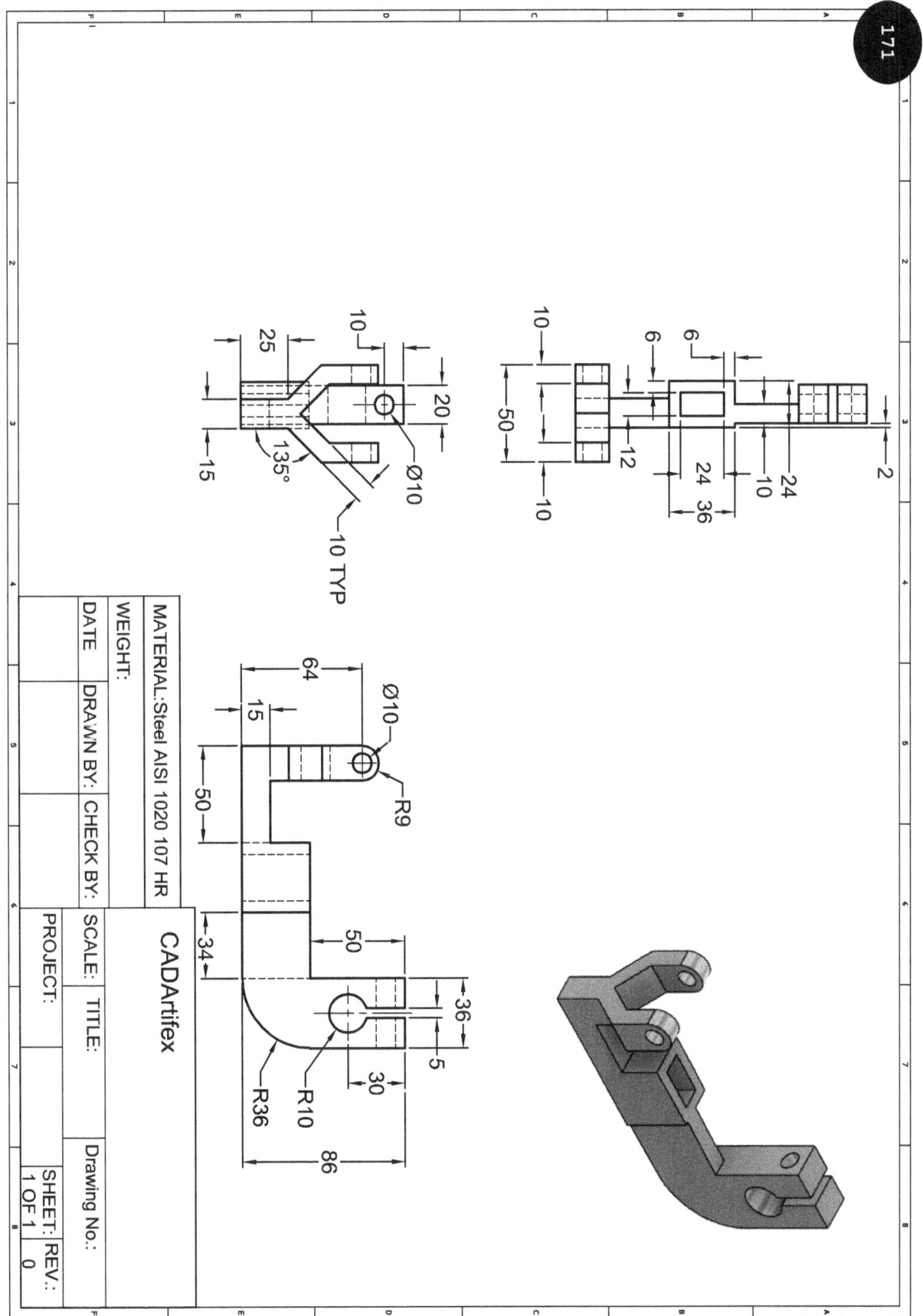

MATERIAL:Steel AISI 1020 107 HR

CADArtifex

Exercise 87.

Create the 3D model, as shown in Figure 172. After creating the model, assign the Steel AISI 1020 107 HR material and calculate its mass properties. All dimensions are in mm.

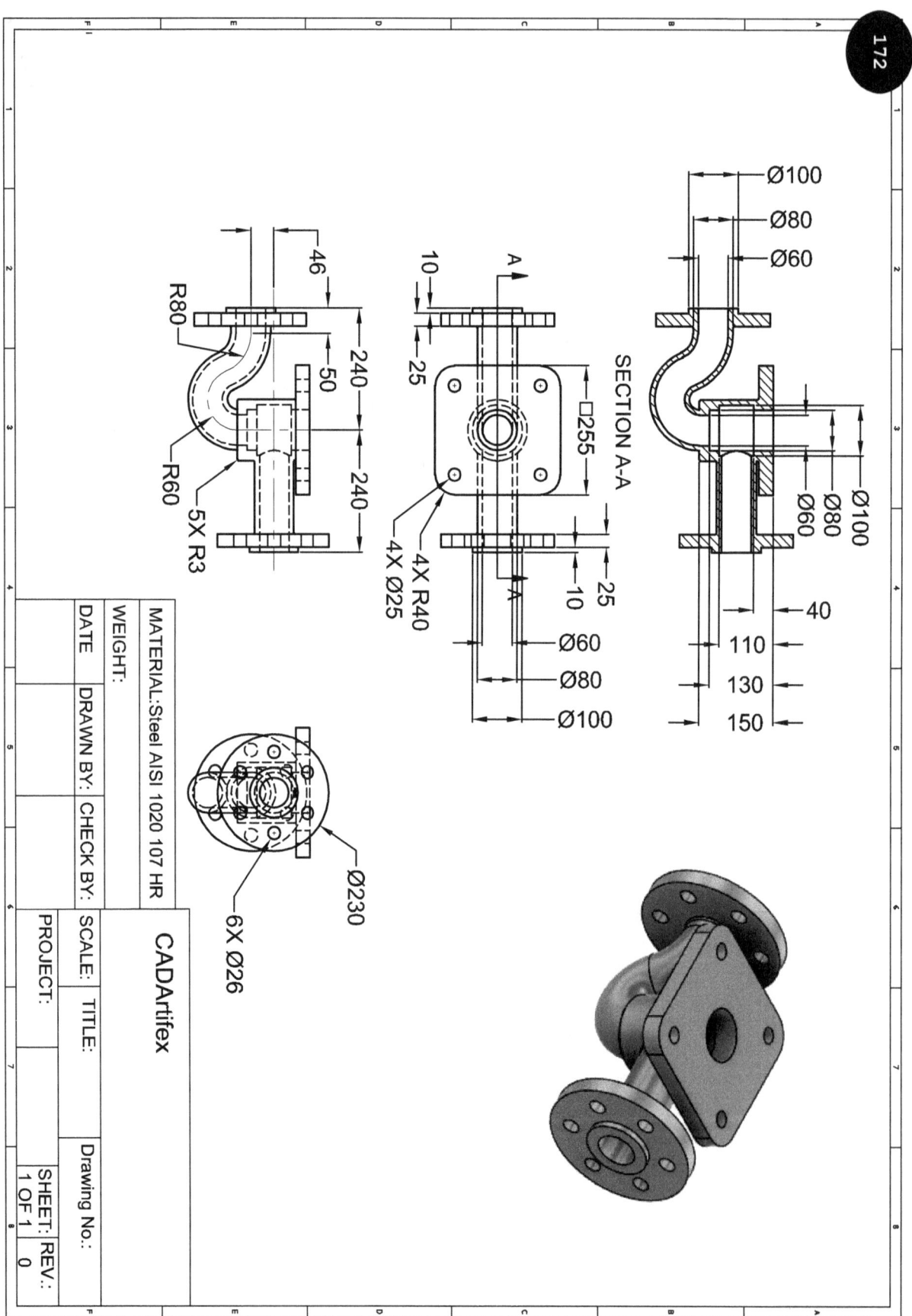

Exercise 88.

Create the 3D model, as shown in Figure 173. After creating the model, assign the Stainless Steel AISI 304 material and calculate its mass properties. All dimensions are in mm.

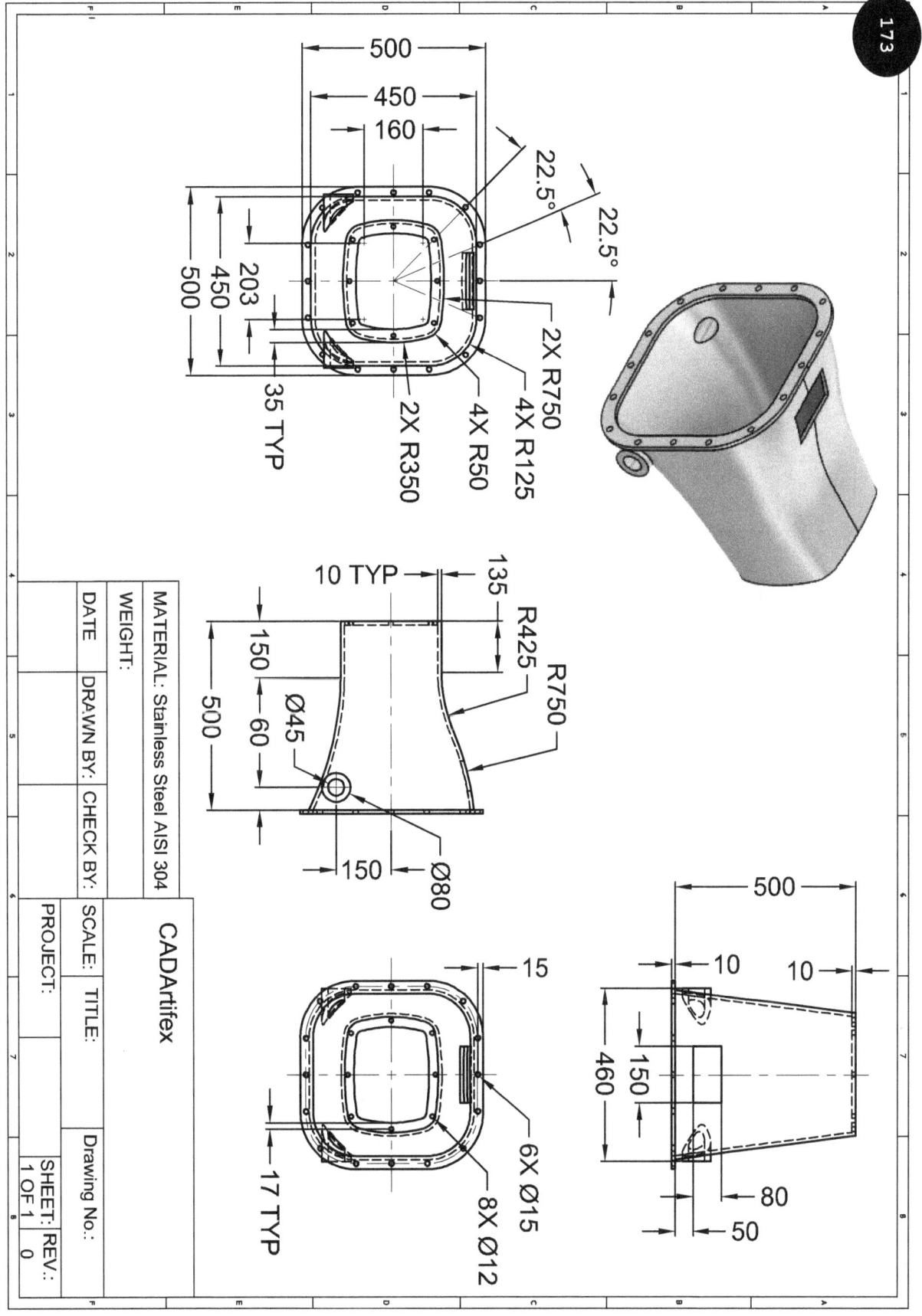

Exercise 89.

Create the 3D model, as shown in Figure 174. After creating the model, assign the Steel AISI 1020 107 HR material and calculate its mass properties. All dimensions are in mm.

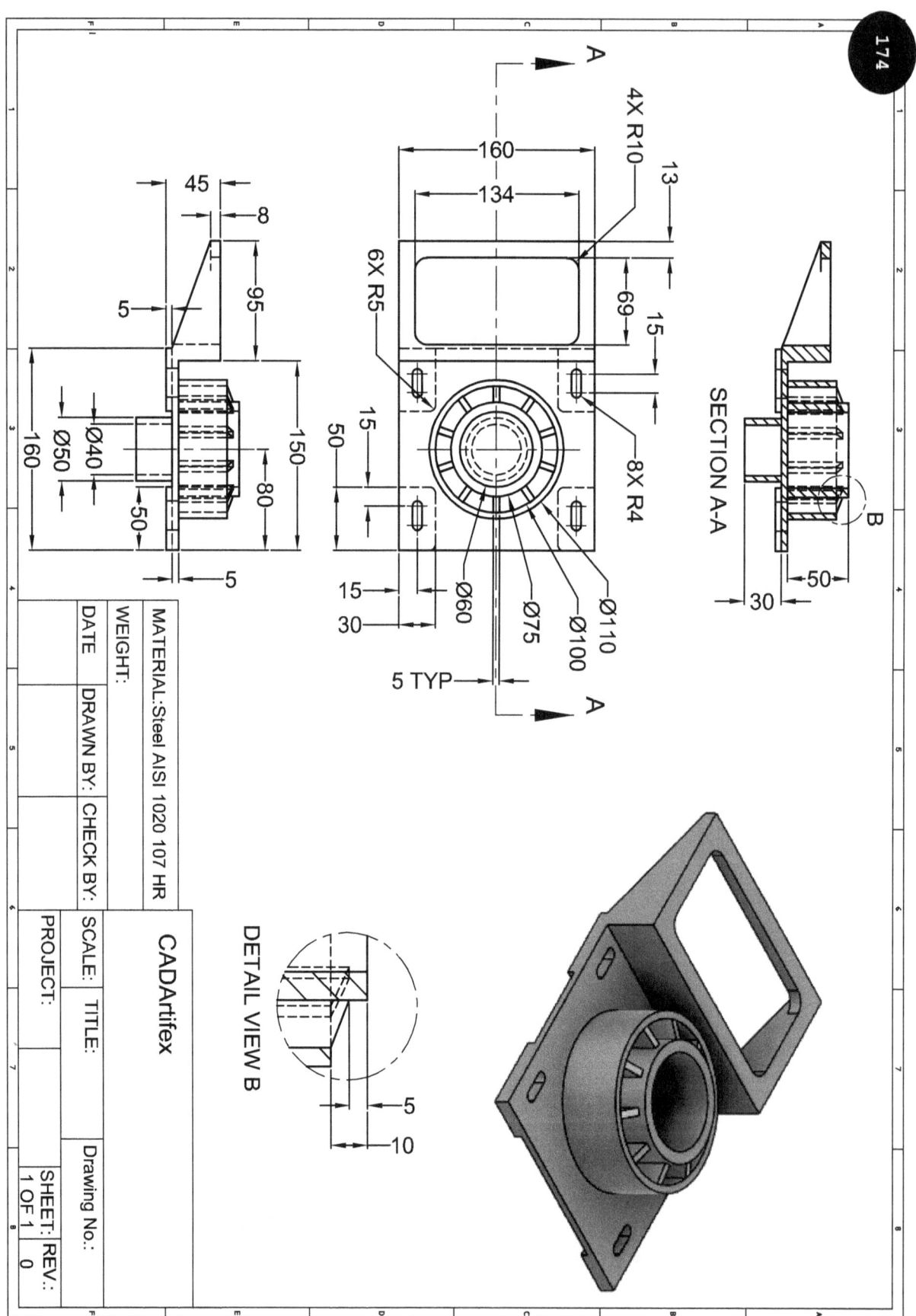

Exercise 90.

Create the 3D model, as shown in Figure 175. After creating the model, assign the Steel AISI 1020 107 HR material and calculate its mass properties. All dimensions are in mm.

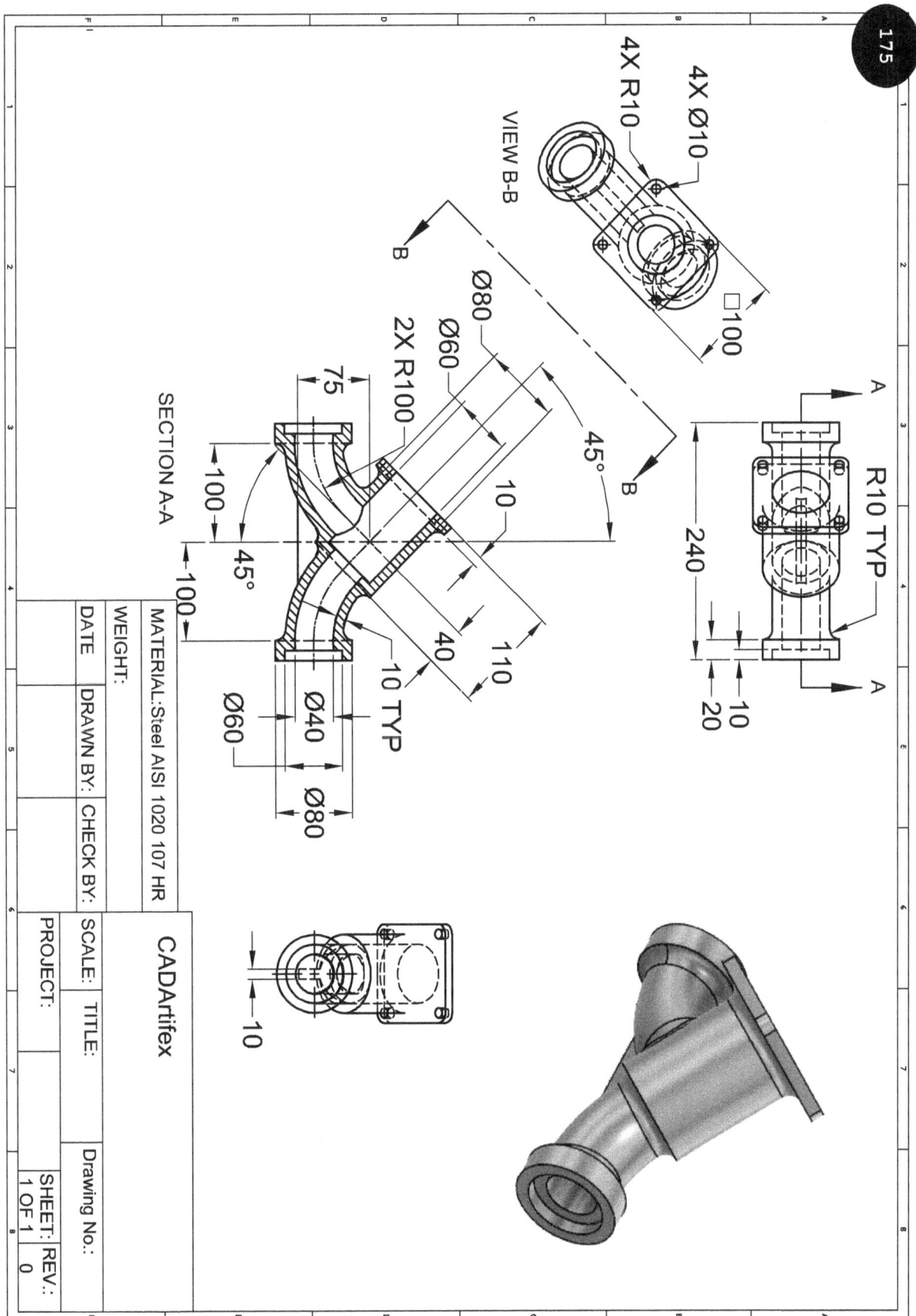

MATERIAL: Steel AISI 1020 107 HR

CADArtifex

Exercise 91.

Create the 3D model, as shown in Figure 176. After creating the model, assign the Platinum material and calculate its mass properties. All dimensions are in mm.

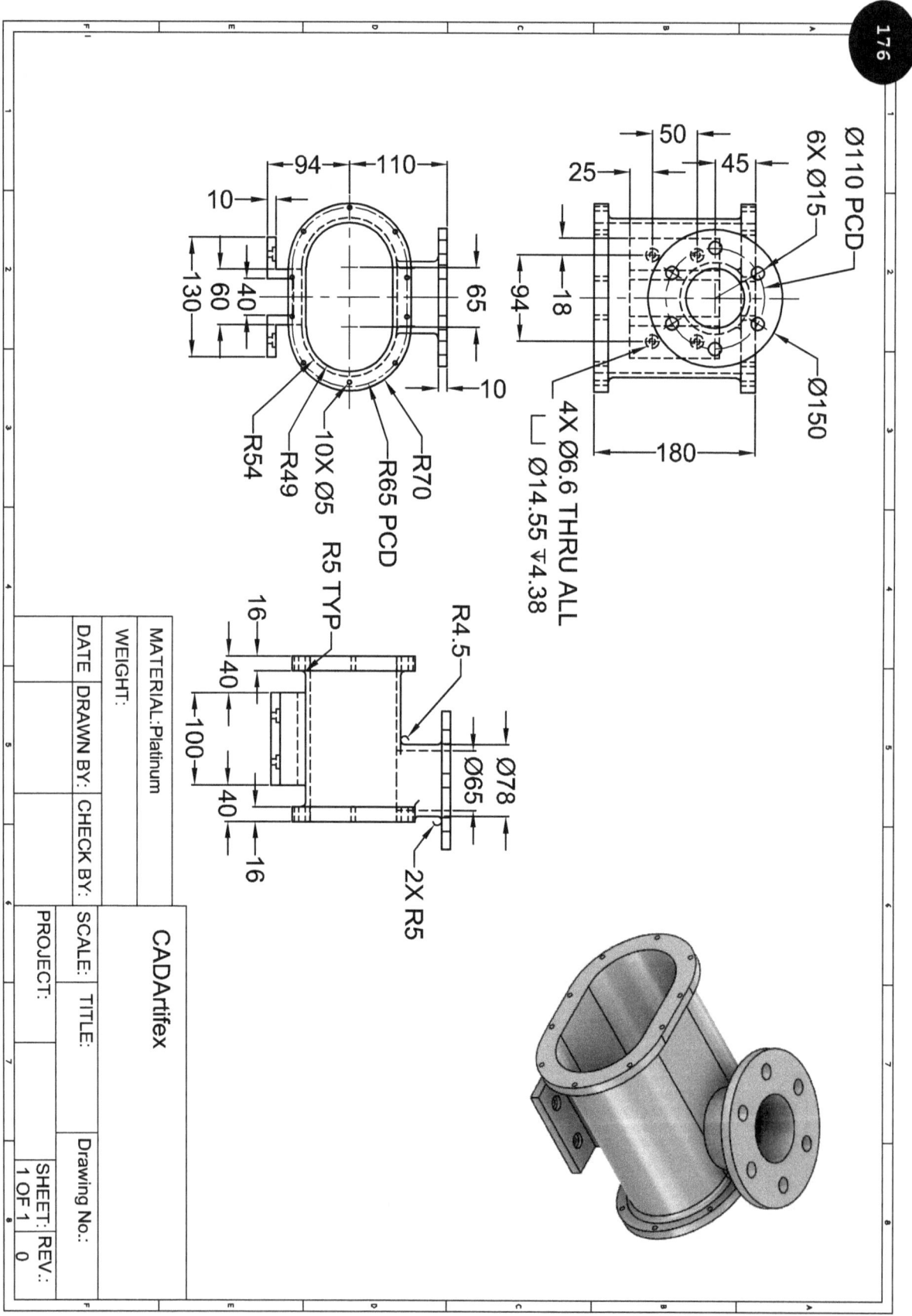

Exercise 92.

Create the 3D model, as shown in Figure 177. After creating the model, assign the Steel AISI 1020 107 HR material and calculate its mass properties. All dimensions are in mm.

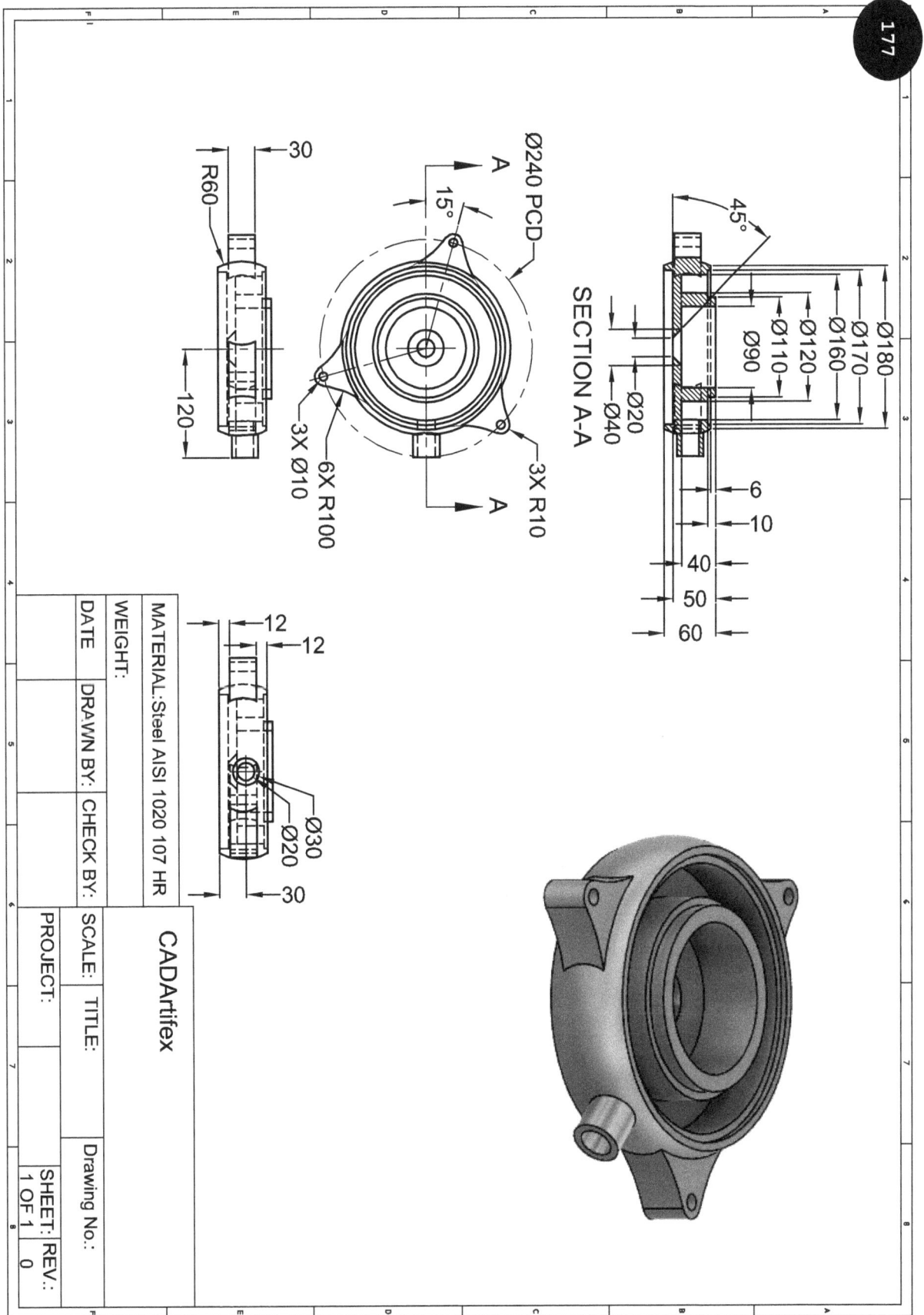

Exercise 93.

Create the 3D model, as shown in Figure 178. After creating the model, assign the Platinum material and calculate its mass properties. All dimensions are in mm.

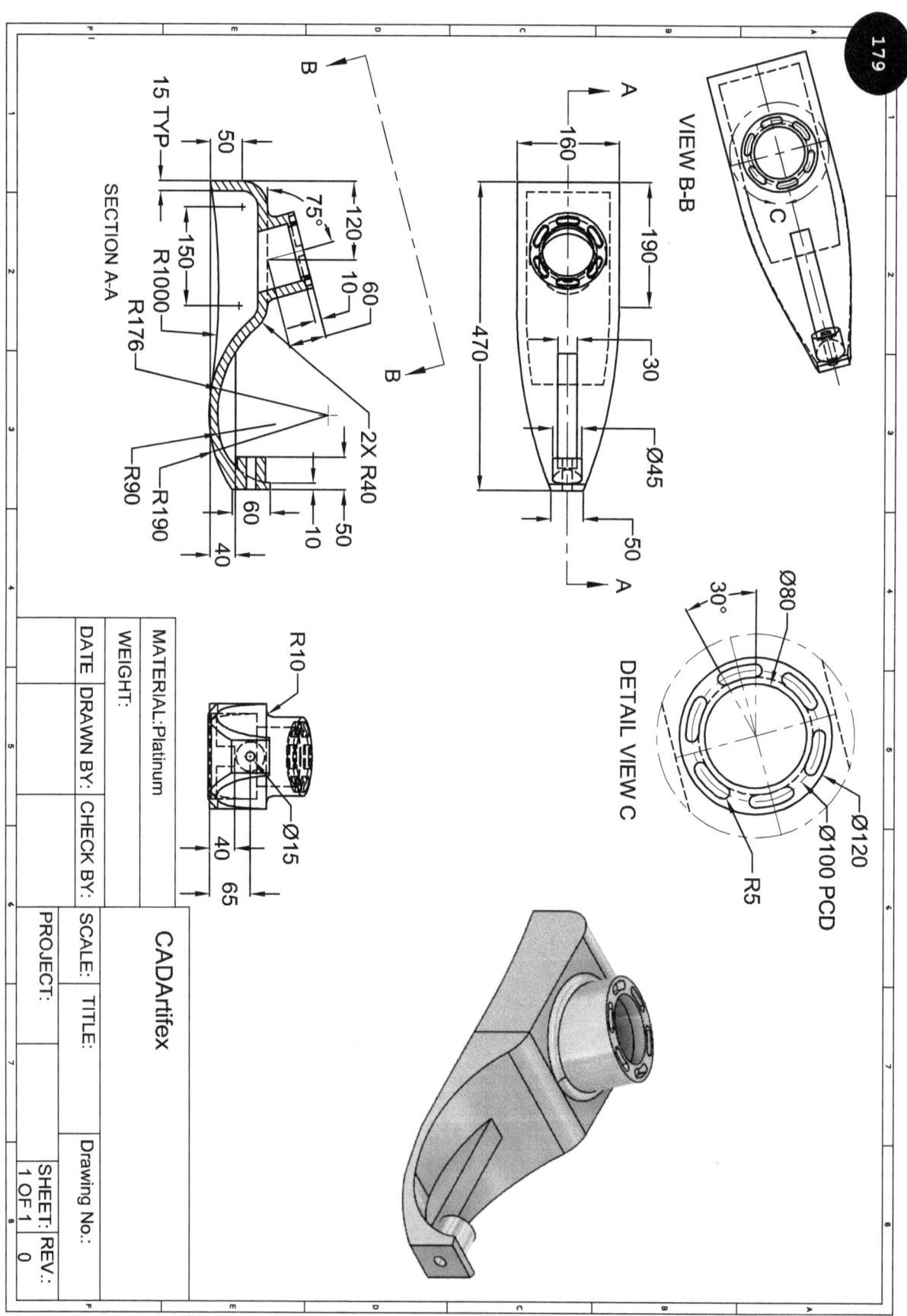

Exercise 94.

Create the 3D model, as shown in Figure 180. After creating the model, assign the Stainless Steel material and calculate its mass properties. All dimensions are in mm.

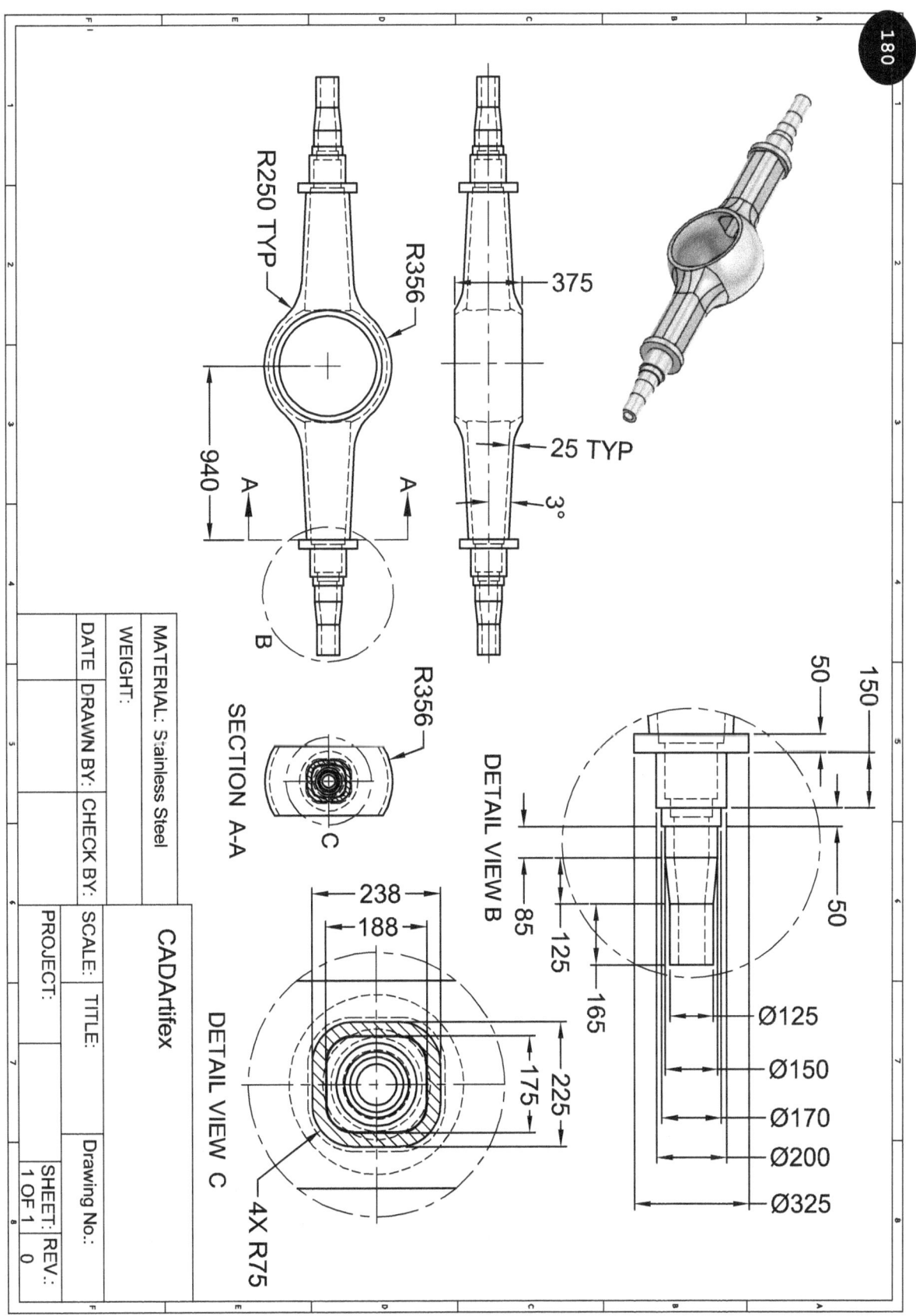

Exercise 95.

Create the 3D model, as shown in Figure 181. After creating the model, assign the Platinum material and calculate its mass properties. All dimensions are in mm.

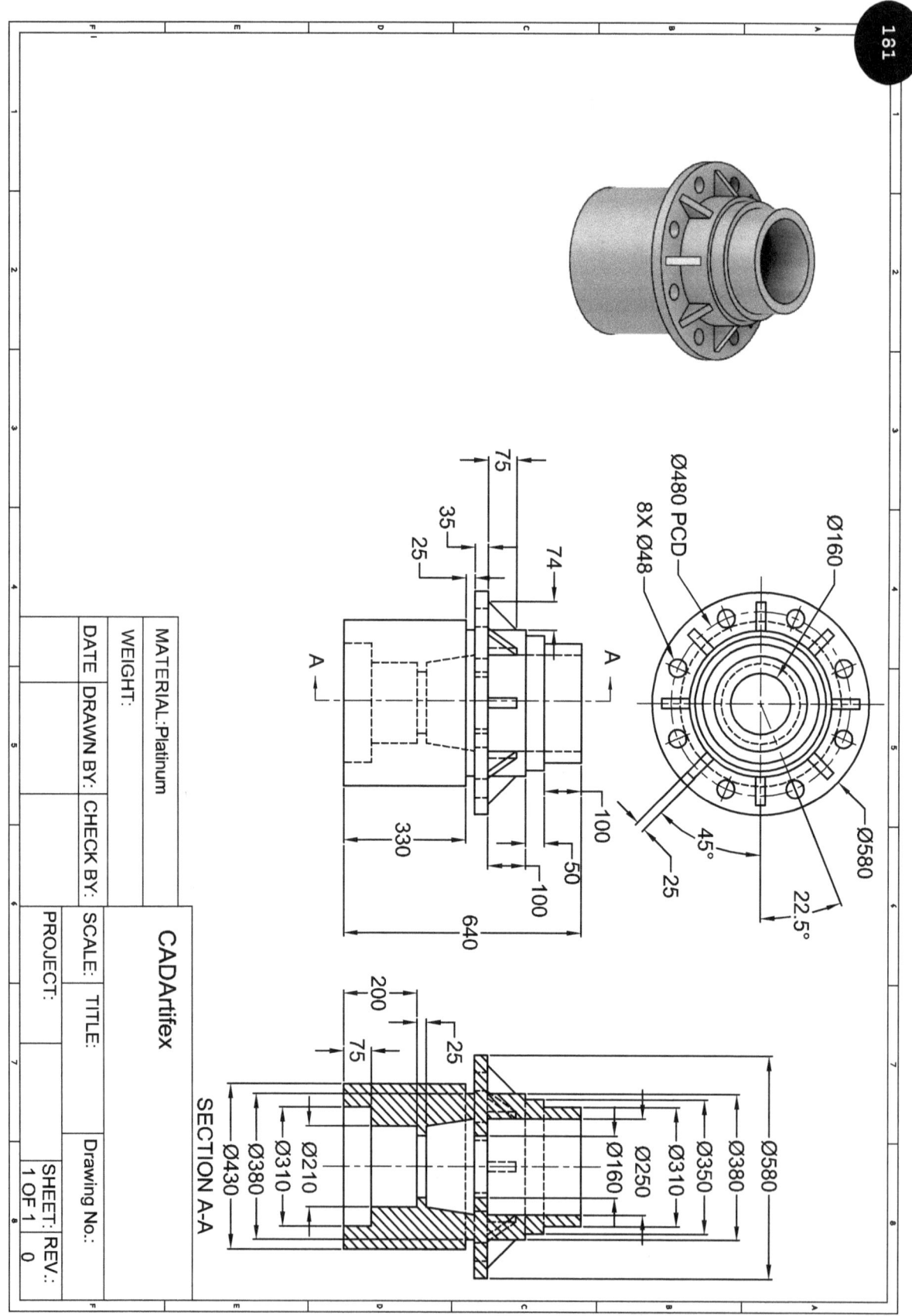

Exercise 96.

Create the 3D model, as shown in Figure 182. After creating the model, assign the Platinum material and calculate its mass properties. All dimensions are in mm.

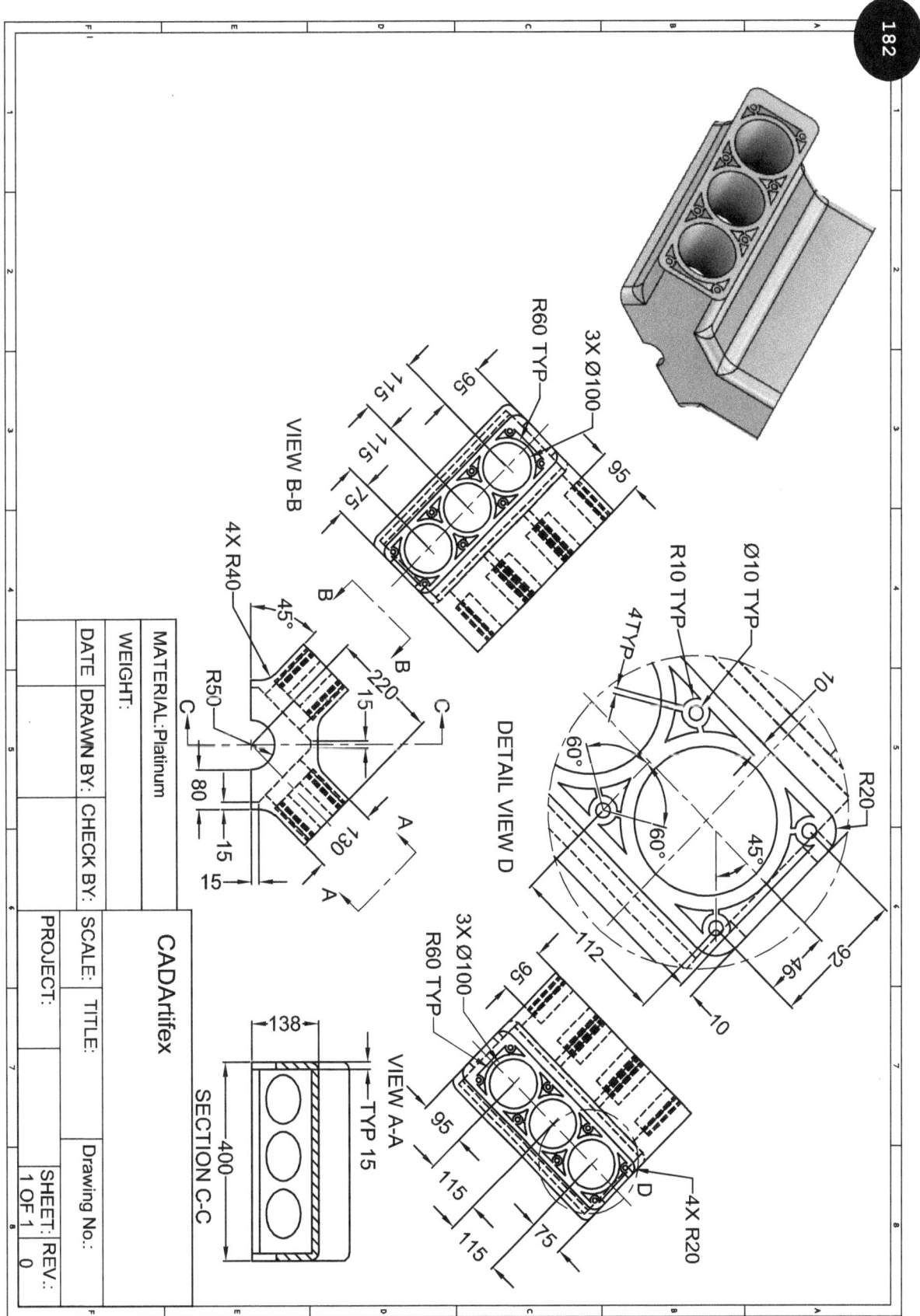

Exercise 97.

Create the 3D model, as shown in Figure 183. After creating the model, assign the Platinum material and calculate its mass properties. All dimensions are in mm.

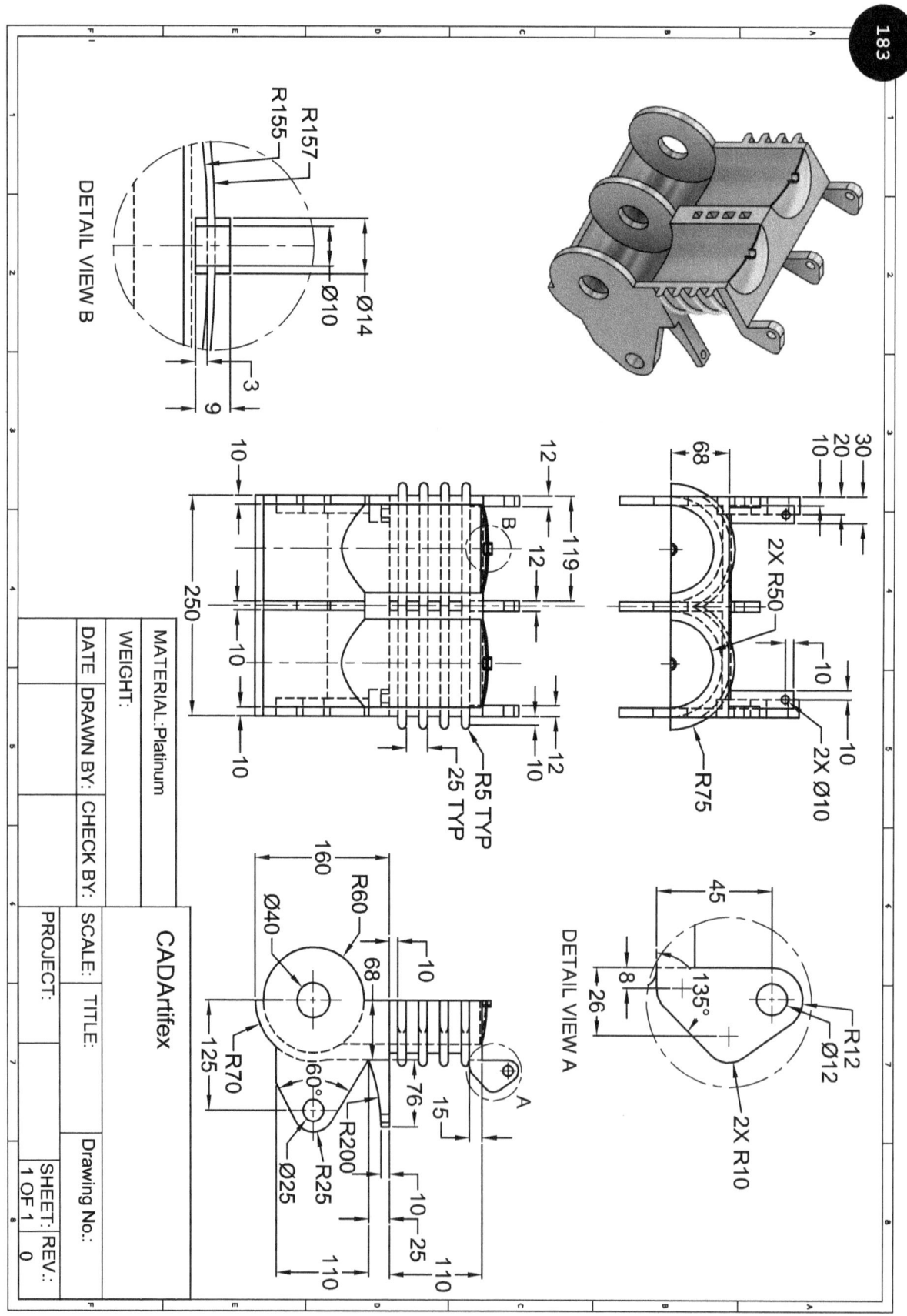

Exercise 98.

Create the 3D model, as shown in Figure 184. After creating the model, assign the Steel AISI 1020 107 HR material and calculate its mass properties. All dimensions are in mm.

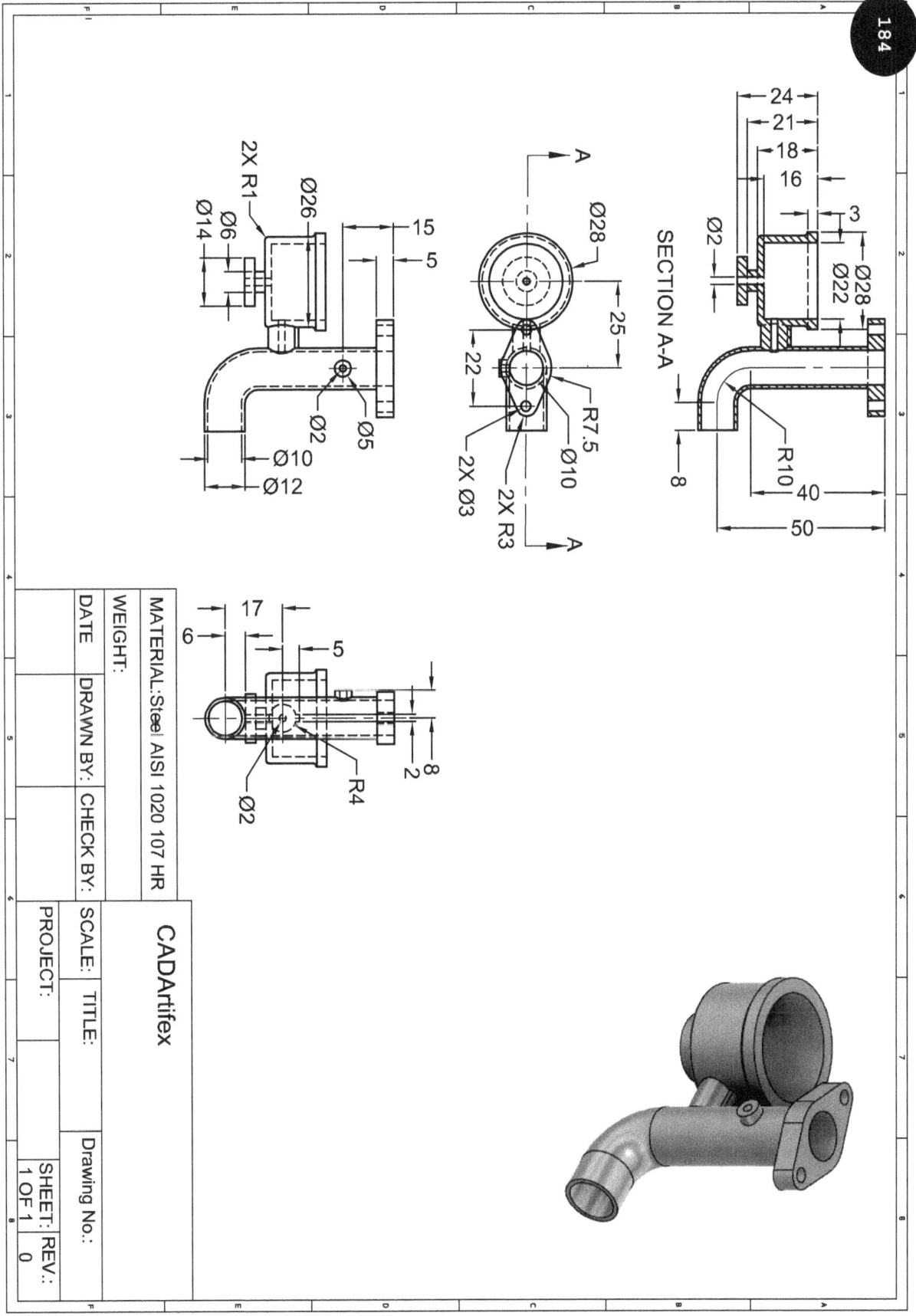

Exercise 99.

Create the 3D model, as shown in Figure 185. Different views of the model and dimensions are shown in Figure 186. After creating the model, assign the Stainless Steel AISI 304 material and calculate its mass properties. All dimensions are in inches.

185

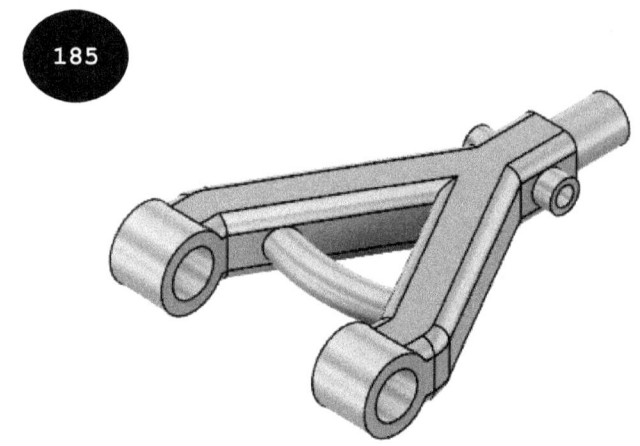

188

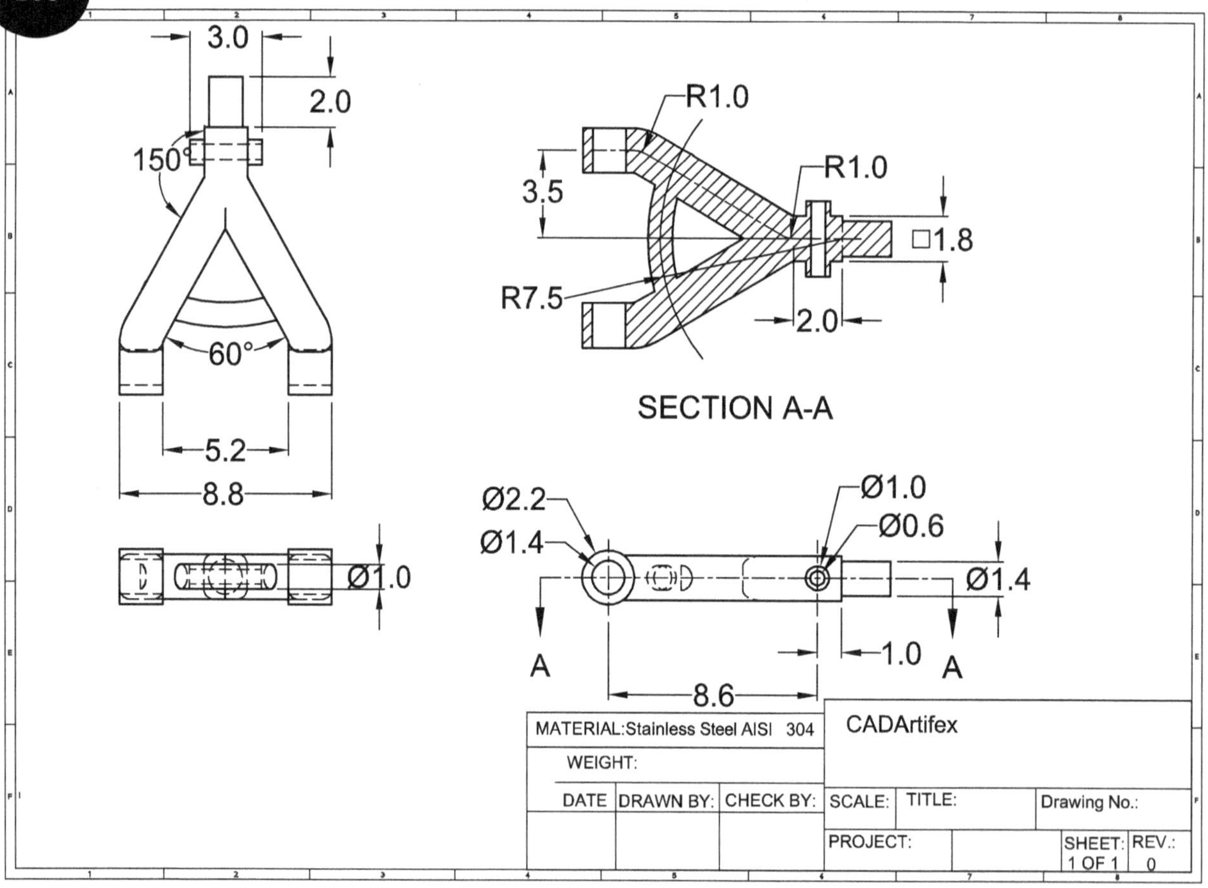

R1.0

R1.0

3.5

□1.8

R7.5

2.0

SECTION A-A

3.0

2.0

150°

60°

5.2

8.8

Ø1.0

Ø2.2

Ø1.4

Ø1.0

Ø0.6

Ø1.4

1.0

A

A

8.6

MATERIAL: Stainless Steel AISI 304			CADArtifex		
WEIGHT:					
DATE	DRAWN BY:	CHECK BY:	SCALE:	TITLE:	Drawing No.:
			PROJECT:		SHEET: 1 OF 1 / REV.: 0

Exercise 100.

Create the 3D model, as shown in Figure 187. Different views of the model and dimensions are shown in Figure 188. After creating the model, assign the Stainless Steel material and calculate its mass properties. All dimensions are in mm.

187

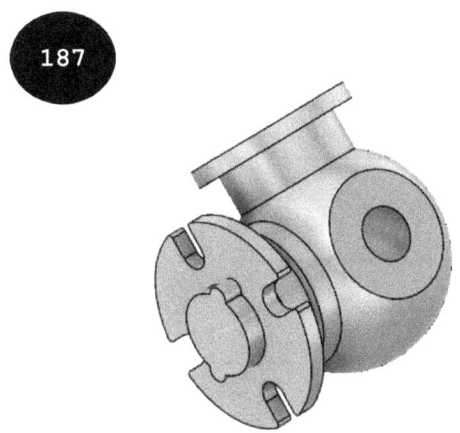

186

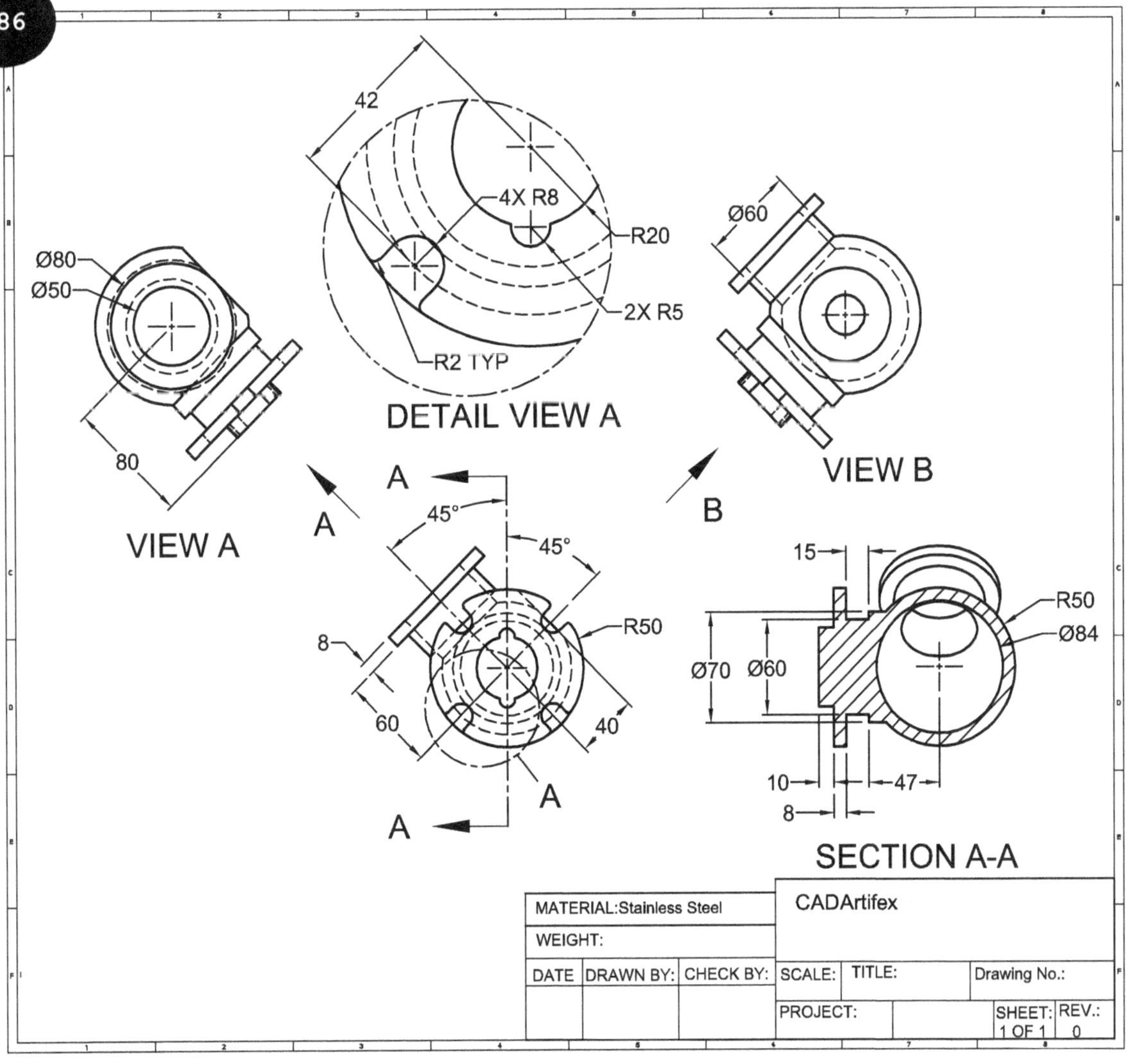

DETAIL VIEW A

VIEW A

Ø80
Ø50

80

42

4X R8

R20

2X R5

R2 TYP

Ø60

VIEW B

A

45°
45°

8

60

40

R50

A

A

B

15

Ø70 Ø60

10

8

47

R50
Ø84

SECTION A-A

MATERIAL:Stainless Steel		CADArtifex		
WEIGHT:				
DATE	DRAWN BY:	CHECK BY:	SCALE:	TITLE:		Drawing No.:
			PROJECT:			SHEET: REV.: 1 OF 1 0

Mass Properties in Grams of Each Exercise:

Exercise No.	Mass in Grams (g)
1	1168.12
2	148.18
3	127.66
4	44920.42
5	1562.20
6	20099.33
7	3470.28
8	144.62
9	13209.25
10	464.05
11	3280.27
12	37267.95
13	15958.71
14	12344.46
15	210.14
16	176.70
17	75.39
18	12198.07
19	318.81
20	2270.40
21	21.78
22	427.95
23	104.36
24	5203.15
25	6618.76
26	1145.29
27	250.64
28	14.53
29	5750.76
30	6553.73
31	620.83
32	2183.84
33	831.77

34	31739.16
35	144.91
36	1397.70
37	1787.96
38	1794.91
39	608.37
40	3959.14
41	1803.31
42	1301.55
43	432072.34
44	21517.82
45	7173.80
46	845.14
47	2644.13
48	71330.53
49	287.49
50	234.22
51	3775.51
52	2206.49
53	816.86
54	814.57
55	14536.85
56	830.56
57	4732.44
58	1952.21
59	13675.53
60	3864.17
61	189.44
62	16144.16
63	15250.33
64	2276.47
65	69.00
66	140.43
67	18738.74
68	20709.03
69	285.58

70	3690.94
71	16978.83
72	2227.69
73	2610.76
74	23319.26
75	18357.17
76	30391.97
77	10801.70
78	115519.26
79	129.59
80	1109.03
81	351.17
82	16147.34
83	602.61
84	1178.49
85	2317.63
86	868.95
87	40170.68
88	56947.15
89	3818.29
90	5505.56
91	21250.83
92	6743.60
93	42996.86
94	519604.06
95	1177158.84
96	165131.12
97	45925.33
98	59.46
99	0.51
100	2769.41

Other Publications by CADArtifex
Some of the other Publications by CADArtifex are given below:

AutoCAD Textbooks
AutoCAD 2024: A Power Guide for Beginners and Intermediate Users
AutoCAD 2023: A Power Guide for Beginners and Intermediate Users
AutoCAD 2022: A Power Guide for Beginners and Intermediate Users
AutoCAD 2021: A Power Guide for Beginners and Intermediate Users
AutoCAD 2020: A Power Guide for Beginners and Intermediate Users
AutoCAD 2019: A Power Guide for Beginners and Intermediate Users
AutoCAD 2018: A Power Guide for Beginners and Intermediate Users
AutoCAD 2017: A Power Guide for Beginners and Intermediate Users
AutoCAD 2016: A Power Guide for Beginners and Intermediate Users

AutoCAD For Architectural Design Textbooks
AutoCAD 2023 for Architectural Design: A Power Guide for Beginners and Intermediate Users
AutoCAD 2022 for Architectural Design: A Power Guide for Beginners and Intermediate Users
AutoCAD 2021 for Architectural Design: A Power Guide for Beginners and Intermediate Users
AutoCAD 2020 for Architectural Design: A Power Guide for Beginners and Intermediate Users
AutoCAD 2019 for Architectural Design: A Power Guide for Beginners and Intermediate Users

Autodesk Fusion 360 Textbooks
Autodesk Fusion 360: A Power Guide for Beginners and Intermediate Users (6th Edition)
Autodesk Fusion 360: A Power Guide for Beginners and Intermediate Users (5th Edition)
Autodesk Fusion 360: A Power Guide for Beginners and Intermediate Users (4th Edition)
Autodesk Fusion 360: A Power Guide for Beginners and Intermediate Users (3rd Edition)
Autodesk Fusion 360: A Power Guide for Beginners and Intermediate Users (2nd Edition)
Autodesk Fusion 360: A Power Guide for Beginners and Intermediate Users

Autodesk Fusion 360 Surface and T-Spline Textbooks
Autodesk Fusion 360 Surface Design and Sculpting with T-Spline Surfaces (6th Edition)
Autodesk Fusion 360 Surface Design and Sculpting with T-Spline Surfaces (5th Edition)
Autodesk Fusion 360: Introduction to Surface and T-Spline Modeling

Autodesk Inventor Textbooks
Autodesk Inventor 2024: A Power Guide for Beginners and Intermediate Users
Autodesk Inventor 2023: A Power Guide for Beginners and Intermediate Users
Autodesk Inventor 2022: A Power Guide for Beginners and Intermediate Users
Autodesk Inventor 2021: A Power Guide for Beginners and Intermediate Users
Autodesk Inventor 2020: A Power Guide for Beginners and Intermediate Users

FreeCAD Textbooks
FreeCAD 0.20: A Power Guide for Beginners and Intermediate Users

PTC Creo Parametric Textbooks
Creo Parametric 10.0: A Power Guide for Beginners and Intermediate Users
Creo Parametric 9.0: A Power Guide for Beginners and Intermediate Users
Creo Parametric 8.0: A Power Guide for Beginners and Intermediate Users
Creo Parametric 7.0: A Power Guide for Beginners and Intermediate Users
Creo Parametric 6.0: A Power Guide for Beginners and Intermediate Users
Creo Parametric 5.0: A Power Guide for Beginners and Intermediate Users

SOLIDWORKS Textbooks

SOLIDWORKS 2023: A Power Guide for Beginners and Intermediate User
SOLIDWORKS 2021: A Power Guide for Beginners and Intermediate User
SOLIDWORKS 2020: A Power Guide for Beginners and Intermediate User
SOLIDWORKS 2019: A Power Guide for Beginners and Intermediate User
SOLIDWORKS 2018: A Power Guide for Beginners and Intermediate User
SOLIDWORKS 2017: A Power Guide for Beginners and Intermediate User
SOLIDWORKS 2016: A Power Guide for Beginners and Intermediate User
SOLIDWORKS 2015: A Power Guide for Beginners and Intermediate User

SOLIDWORKS Sheet Metal and Surface Design Textbooks
SOLIDWORKS Sheet Metal and Surface Design 2023
SOLIDWORKS Sheet Metal Design 2022
SOLIDWORKS Surface Design 2021 for Beginners and Intermediate Users

SOLIDWORKS Simulation Textbooks
SOLIDWORKS Simulation 2023: A Power Guide for Beginners and Intermediate User
SOLIDWORKS Simulation 2022: A Power Guide for Beginners and Intermediate User
SOLIDWORKS Simulation 2021: A Power Guide for Beginners and Intermediate User
SOLIDWORKS Simulation 2020: A Power Guide for Beginners and Intermediate User
SOLIDWORKS Simulation 2019: A Power Guide for Beginners and Intermediate User
SOLIDWORKS Simulation 2018: A Power Guide for Beginners and Intermediate User

Exercises Books
Some of the exercises books are given below:

AutoCAD Exercises Books
100 AutoCAD Exercises - Learn by Practicing (2 Edition)
100 AutoCAD Exercises - Learn by Practicing (1 Edition)

Autodesk Inventor Exercises Books
Autodesk Inventor Exercises - Learn by Practicing

SOLIDWORKS Exercises Books
SOLIDWORKS Exercises - Learn by Practicing (3 Edition)
SOLIDWORKS Exercises - Learn by Practicing (2 Edition)

www.ingramcontent.com/pod-product-compliance
Lightning Source LLC
Chambersburg PA
CBHW082135290526
45794CB00008B/3051